Eggshells of Eagles
The Early Days of the Hazardous Waste Industry

Peter J. Gorton

RPSS Publishing - Buffalo, New York

All rights reserved. Copyright ©2023 Peter Gorton

All rights reserved. No part of this publication may be reproduced or distributed in any form or by any means, or stored in a database or retrieval system, without the prior written permission of the publisher.

publisher@rockpapersafetyscissors.com

ISBN:978-1-956688-20-7

Printed in the United States of America

10 9 8 7 6 5 4 3 2 1

RPSS Publishing - Buffalo, New York

Dedication

Two colleagues who have passed away
-John Duwaldt and James B. Moore

My Mom & Dad – John and Blanch Gorton

Brothers John *(who sadly passed in 2023)*
& Robert Gorton

My Daughter Karlee Lewis
and Son Dustin Gorton

Acknowledgements:

My lifelong friends Bill and Andrea Baker
and their two daughters Katrina and Becky
Mark Riforgiat
Dave Dahlstrom
Michael Twardy
And the many others who supported my attempt
to turn a technical subject into a novel.

I would also like to acknowledge the women and men who knowingly, willingly, or through ignorant naivete accepted the risk that put them at greater hazard than they probably knew. It's about a profession that is little known and little written about but like others has its heroes and villains. This story is not meant to idolize or make more of them than their worth but simply to acknowledge them and their profession. Some of them were the best and the brightest in their fields and some are no longer with us. They don't get publicity or visibility working away in private firms or government agencies—unlike the preening stars of the big screen or athletic fields or the so-called "public servants" of the country who treat themselves like noble women and men and pontificate on things they know little about. They have no soap box to stand on, although many should. They have so much to say, and there's so much they have done.

TABLE OF CONTENTS

Prologue	7
CHAPTER 1 Johnston Atoll	13
CHAPTER 2 He Wanted to Be Great	32
CHAPTER 3 I Don't Like Watching You Die Because I Know I Cannot Bring You Back	41
CHAPTER 4 The First Hazardous Waste Training	53
CHAPTER 5 Audits at MIDCO I and MOTCO Too	65
CHAPTER 6 Nick and the Spy	76
CHAPTER 7 Two Sites in O Hi O	85
CHAPTER 8 Indianapolis to Ashtabula, Ohio, and a Visit by JAFOs	93
CHAPTER 9 22 Caliber Killer and the Midtown Slasher	106
CHAPTER 10 Go West, Young Man, and as the Salmon Swim	113
CHAPTER 11 A Senseless Death	122
CHAPTER 12 Petro Processing – Baton Rouge	133
CHAPTER 13 Love Canal and the Pfohl Brothers Landfill	143
CHAPTER 14 Vibrocoring on Alcyon Lake and Sampling Rabbit Run	157
CHAPTER 15 Ode to John and the Jacksonville Oil Pits	165
CHAPTER 16 The Eggshells of Eagles -from DDT to PFAS	173
CHAPTER 17 "Please Control Your Soul's Desire for Freedom"	179
Appendix – Additional Technical Detail	195
References	199
About the Author	206

Prologue

Thinking back, letting the memories roll across your emotions in bittersweet bits of your history never to be captured anywhere ever again; unique memories that you share with only a few others so special in time and captured in your essence only. There are times in your life when you actually take the time to reflect on things you're never going to do again, witness again, or have the joy or sorrow of again except, perhaps, if you allow the memory to be captured in photographs or the written word. Like never watching again the unique interplay of your late wife with your kids. Or the memory of walking between your mom and dad's house to your brothers in the dark of a starlit summer night in Cape Cod, taking in the beauty of the magnificent sky unencumbered by urban light. Likewise he memories as a young kid of your first trip to the city or your first trip to the country. Or the 4th of July party that your parents and neighbors held every year between their Evergreen Knolls, Town of Cortlandt, New York backyards that always devolved into the parents singing their favorite songs from yesteryear and the kids running around with sparklers or playing some other games while the keg of beer dwindled along with the night. The ice chunks you watched created with the pick on the big block of square ice in the early morning dew long ago melted along with the daylight. The Castagnettas, the Cardinales, Sheridans, Oswalds, Bakers, Hyslops, and the other families that shared moments in time in their backyards never to be recaptured again. Those events, although surely different in various ways, happen in each and every person's life.

Interconnections sometimes run parallel before turning perpendicularly and intersecting our lives. Take waste for example. Humans produce waste, and the type of wastes we produce affects many things—most importantly ourselves. A question, however, is will these wastes "get us" in the end. Will they interconnect somewhere, somehow?

Some wastes are toxic or hazardous, but even so some waste, maybe most, can be used for useful purposes by those with enlightened minds. We do know this: a toxic substance is not always hazardous, and a hazardous substance may not be toxic. Nuances such as these are meaningless to most people in their everyday lives. Yet people interact with toxic and hazardous substances daily, and for the most part they deftly or ignorantly maneuver past them. Or do they? Pretty much everything in our everyday world is hazardous; it's a matter of exposure and dose as to their toxicity. Water is toxic if we drink too much of it. This type of intoxication—also known as water poisoning, overhydration, or water toxemia—can result in a fatal disturbance in brain functions due to the normal balance of electrolytes in the body being pushed outside safe limits by excessive water intake. What other things in life affect our normal balance, pushing our safe limits beyond their acceptable levels?

Something can be extremely toxic but not at all hazardous if there is no exposure or exposure is of de minimis risk. Then there is the whole chronic versus acute exposure or, worse, the effects are exacerbated by other adjunctive risk factors. It is well known, for instance, that in the workplace, workers who are simultaneously exposed to a combination of harmful stressors—such as physical, chemical, biological, and psychological agents—will be at higher risk of health consequences. We all maneuver through stress in our daily lives knowingly or unknowingly contemplating the risk and rewards; some claim ignorance and victimhood later.

When we think of a toxic substance or waste, we typically relate this to something being poisonous and dangerous that will probably cause immediate harm to plants, animals, and humans. Toxic substances, however, are not hazardous unless there is a route of exposure, a sufficient dose or potency, and an adequate length of exposure. Most people stay far away from things that are known to be "toxic" or "hazardous." However, early hazardous waste workers were educated, paid, and trained to walk towards them and as such in my opinion should have limited to no claim of victimhood.

Innovation drives the future of our population. But populations keep increasing, and demand doesn't stop. Some ignorantly think we can innovate with no effect on the environment. Others rightfully understand the key is to reduce the effects since innovation will always likely continue. Some believe if we looked at human population as we do animal populations when they get too large and dangerous, we would cull the herd. There is a lot of talk today about building a safer, more sustainable world. But this is no easy task, especially when the motivation of the corrupt is something other than science and the good of mankind.

It appears easy to deceive many people—if not a large portion of the population—at any one time, especially if heart strings are pulled or when the population is purposefully frightened. We have seen this through history as the carnival barkers promise things they don't even understand or believe as they sell that cure-all bottle of tonic to the gullible crowd. It's worse when that gullible crowd is made more pliable using fear for their wellbeing or by using "religious" overtones parsed by people we are supposed to trust—teachers, leaders, politicians—or people we falsely emulate, such as actors and sports heroes. Today we see the same old ancient play in the "religion of green," and its consequential effects will ultimately be the same. Saving the planet is a worthy goal, for sure. Who could argue with that? Killing the messenger or corrupting the message of science is not.

It's very disconcerting when you realize that a few play the many and seem to do it so very easily. This was never truer than during the recent pandemic, and you would think the lesson would be evident. But it never is, is it? You would think that a simple dose of logic wrapped around commonsense intelligence should be all that is needed to see through the quackery. Sure, making your life's ambition one that cares about the environment or human health and wellbeing is more than a noble cause or a worthy profession. Using that energy to inflict pain and suffering under a false guise or phony science is not. Do the ends ever justify the means? That people are so easily led and so easily fooled is both disheartening and infuriating because it is caused by evil contempt of science, math, physics—and, in the end, human need. Once the evil behind the concepts make believers out of the "noble masses," they manipulate them in such demonic, enthusiastic ways that even the most obvious oxymoronic concepts become acceptable in perfunctory ways.

In 1976, the Resource Conservation and Recovery Act (RCRA) was enacted. It's not that humans had not produced waste before 1976 and throughout their existence. It's just that prior to 1976, these wastes were dumped into the nearest hole, marsh, river, lake, or ocean or onto an undesirable or underused piece of land. Many times, this was done by people who really should have known better. The concept of toxicity and hazard was certainly not new in 1976. Oh sure, the axioms "out of sight, out of mind" or "the solution to pollution is dilution" were at play. But some were just not informed or educated enough to really know or even suspect the truth, so they bought into these flawed concepts. Many years later, their descendants would uncover the deed and pay some price. But even the noble cause of cleaning waste was usurped by corrupt humans. And whole industries and government agencies grew into overlapping absurd bureaucracies. Some made money studying things to death and never coming to a solution—or at least never coming to one in an efficient and economic manner.

Over time, it became clear to the general population that an "out of sight, out of mind" strategy for waste management was, in fact, neither. Of course, true to history, it typically necessitated some shock-inducing catastrophe or publicized event to motivate our population to action. Such events as those at Love Canal or Times Beach, Missouri; the WR Grace site in Woburn, Massachusetts; or the Valley of the Drums site near Louisville, Kentucky; and the even more visually catastrophic Bhopal disaster in India are all examples. The question is: Who was doing the motivation, and was it in a noble direction?

It became clear that some wastes posed greater hazards than others, and those hazards required the proper response—perhaps even from the government. Two of the most notorious incidents that brought the required publicity to effectuate a response were Love Canal in Niagara Falls, New York, and Times Beach seventeen miles west of St. Louis. These are still notorious today, as is the Woburn, Massachusetts, site made more infamous by the bestselling book A Civil Action by Jonathan Harr and a movie of the same name.

It is ironic that some of the same people who got rich causing much of the contamination were the same ones who got rich cleaning it up. Yellow iron is funny that way as is the expertise and wherewithal to use it. Also, there is something to be said for knowing where the bodies are buried.

The following tells the story of the early hazardous waste workers through the experiences of one of them, Nick. These are reflections of times and unique events that will never happen again—or will they in some alternate way, as history always appears to repeat itself? Only the degree is altered in some significant or insignificant way. Perhaps the history portrayed within will spark the memories in others as life and human experience marches on. When Nick worked on a garbage truck all through undergraduate and graduate school, his boss, knowing his major, would take great glee in declaring how he was putting Nick through school so he could come back and close him down. The irony was never lost on Nick either, and it helped shape much of his beliefs as his career and life wound through the many experiences that shaped his being.

Every year, the U.S. produces over 400 million tons of hazardous waste that is toxic, corrosive, flammable, or explosive. Then there is radiological waste and biological waste. Back in the day, this waste was sometimes just dumped anywhere, despite its threat to public and environmental health. This story is based on true events about the early hazardous waste workers whose mission was to identify, quantify, and in some cases discover these "lost" hazardous and toxic waste dumps across America. Much like the discovery of some ancient Mayan ruin, these sites had in many cases been reclaimed by the "jungle"—be it urban or rural. The extent of their presence had to be meticulously uncovered even if some were right under our noses and we passed by or walked over them daily. The end game, one hopes, was collecting enough data for their safe removal and remedial redesign to help purify the planet. This story touches on a small portion of the events and sites, some of which were famous, and people—who were not—at the beginning of the hazardous waste industry. The early sites across the country are covered in some scientific detail as a nod to the nerds who need that detail and as a backdrop to everyday events, some of which are notorious, and the not-so-glamorous activities of those early workers; fleeting memories captured in the stories that follow.

It's a story to document those who toiled anonymously early in an industry that was designed to help America move on from the mistakes of its industrial past made either though greed, ignorance, or both and into our industrial future. These were environmental professionals—scientists and engineers, chemists and biologists—who worked tirelessly many times in the frigid cold of the northern winters and the steaming oppressive heat of the southern summers in all manner of gear meant to protect them from chemical or radiological hazards or from biological agents and from the physical hazards all around them.

This story is dedicated to the women and men who knowingly, willingly, or through ignorant naivete accepted the risk that put them at greater risk than they probably knew. It's about a profession that is little known and little written about but like others has its heroes and villains. This story is not meant to idolize or make more of them than their worth but simply to acknowledge them and their profession. Some of them were the best and the brightest in their fields. They don't get publicity or visibility working away in private firms or government agencies for adequate wages—unlike the preening stars of the big screen or athletic fields or the so-called "public servants" of the country who treat themselves like noble women and men and pontificate on things they know little about. They have no soap box to stand on, although many should. They have so much to say, and there's so much they have done.

Chapter 1

Johnston Atoll

Nick glanced over his shoulder as something caught his attention behind him. He was holding the metal stake as his partner was pounding it into the crushed coral at the designated grid location they had just marked. Later, after calm reflection about the day he thought perhaps he would die a horrible death at such a young age, he started to second-guess the mission. Nick had started the day in the mess hall eating scrambled eggs and sausage.

It began unceremoniously sometime around 3 or 4 am that morning when he woke in a panic to the overwhelmingly loud engines of the MAC flight landing on the tarmac, probably 100 yards from his barracks. The smell of jet fuel wafting through the open window, a smell that reminded him of the smell from trucks full of petroleum-impacted soil as they rolled by him on a cleanup site he worked a year before. He had little time for nostalgic recollections because they had to be at mess and off to the job site in short order.

Only on the island less than a week, they were halfway through their task of laying out the sampling grid across the former agent orange storage area at the most downwind sector of the island. Certainly not at all used to the noise of a plane landing next to his bed, Nick's first thought was that the island was exploding, and his immediate action was to reach for his military issued chemical mask before realizing it was just the plane. He and his teammate Jake were only a few days on the island, and the extreme nature of things had not yet become casual. Weeks later, he would sleep through the MAC flight landing that came from Hawaii once a week.

For nearly 70 years, Johnston Atoll was under the control of the American military. At that time, it was used as a bird sanctuary, a naval refueling depot, an airbase, for nuclear and biological weapons testing, for space recovery, a secret missile base, and as a chemical weapon and Agent Orange storage and disposal site. These activities left the area environmentally contaminated and extremely dangerous when downwind of the bunkers storing the nerve agents and chemical weapons or when near the burn pit that was used to destroy the defoliant. Nick's mission was to sample the former Agent Orange storage area located at the farthest downwind northwest corner of the island. This was about an acre-sized area where over 30,000 drums of herbicide orange had been stored after the Vietnam War. The drums were gone now, and their task was to sample the vacant area on the far, lonely furthest downwind area with no prior knowledge of what the area looked like when the drums were there. All they saw was an open area of crushed coral and no vegetation.

A few days prior, Nick had caught that telltale herbicide smell as he walked near the former Agent Orange storage area in what he thought was an upwind position. The wind varied slightly, and that was all it took. He knew that smelling the odors signified an exposure to Agent Orange, albeit very minor. Still, he consciously filed this away for consideration. As slight and quick as it was, it was nonetheless heavy and sickly, but Nick figured since he moved away rapidly, his exposure was minimal. He would not make that mistake again, nor would his crew. The wind blew strongly on the island in one direction 99% of the time, and the Agent Orange area was the farthest downwind part of the island—even farther downwind than the biological and chemical warfare bunkers identified as off-limit areas on the base map

He had smelled a similar odor before in recently sprayed orchards and when walking along train tracks that had just been sprayed. As kids, he and the neighborhood children would run after the "fog" truck spraying mosquitoes in their backyards in the sixties, and there was that same smell. Or was it more recently on that vacant, rundown former pesticide plant he was on in Rock Creek, Ohio? A somewhat aromatic, pungent odor with a distinct, slightly phenolic, ether-like chemical smell or some variation of those similar smells. Whatever its description, Nick knew it was the smell of the breakdown products of the 2,4-D and 2,4,5-T in the herbicide and the telltale sign that they were in the right spot for the project mission of sampling for levels of Dioxin in the crushed coral. He would have to extend the perimeter of their work zone to ensure that didn't happen again, especially when they were actively digging test pits later in the project.

On this particular day, as they worked to establish the sampling grid, Nick glanced back and saw an Army Chemical Corp Military Policeman (MP) in full chemical gear motioning him to come quickly. Nick tapped his partner on the shoulder and said, "We got to go." His partner started to say it was not time, and he wanted to finish. Nick cut him off in mid-sentence; they had already argued a few times over the mission and about how to get it done.

"Look, you stay here if you want, but that MP over there wants us now, and I am not waiting to argue with you." Nick stood and started at a full-out run in full protective gear toward the pickup truck the MP was motioning toward. As they came within about 10 feet, the MP yelled, "Get in the back bed," and Nick jumped up and in in a single bound. His partner, Jake, was right on his heels. As soon as they landed in the truck, it stared down the access road at a decent clip and then suddenly stopped abruptly at a checkpoint just barely outside their work zone.

They were facing two armed MPs dressed out in full gear. The full gear looked similar to the mask they had been issued at entry as they exited the plane a few days before. It looked like it was left over from Vietnam. The MPs' weapons were positioned across their chests in what looked like the ready position, which added to the surreal, ominous feel. The military gear was supposed to be designed to cover the skin, reducing the chances of a chemical or nerve agent entering the body through cuts and scratches. It was supposed to protect US soldiers against nuclear, biological, and chemical warfare when worn properly, along with their masks. Nick knew that the protection he and Jake were wearing was superior because of his excessive training and experience in personnel protective equipment. However, that was not making this situation any less charged as he stared at the military weapons and their two stoic soldiers.

Johnston Atoll, also known as Kalama Atoll to Native Hawaiians, is located in the Pacific Ocean, about 717 nautical miles west-southwest of Hawaii. It is an unincorporated territory of the United States and is one of the world's oldest and most remote atolls. Since the base's close, access to the atoll today is restricted to ocean vessels. Johnston Island is the largest of four islands in the atoll complex. The other three islands are Sand Island, a natural islet, and North (Akau) and East (Hikina) Islands, which are man-made. Johnston Island and the three small islets, along with the reef and lagoon, make up Johnston Atoll.

Like most of the over 400 atolls worldwide, Johnston Atoll is located in the Pacific Ocean and is a variation of the typical ring-shaped coral reef that partially or completely encircles a lagoon. Although discovered before by others, it is named after Captain Charles James Johnston, a British sea captain. In 1926, Johnston Island came under the control and jurisdiction of the US Department of Agriculture as a breeding ground and refuge for native birds. In the 1930s, the US initiated its use as a military base. Today, as an unincorporated territory of the United States, it is administered by the US Fish and Wildlife Service of the Department of the Interior as part of the United States Pacific Island Wildlife Refuges. It's a little over seven feet above sea level and contains low-growing vegetation on mostly flat crushed coral terrain.

During Nick's time on the island, potable water used for drinking and bathing was supplied by the island's desalination plant because there are no natural freshwater resources. When Nick was on the island, the partially man-made island was only 2½ miles long and a half-mile wide – just one giant runway with barracks and support facilities on either side.

The United States Air Force (USAF) administered the island when Nick was there, but many of the service men were part of the Army Chemical core. The "Dragons of the Battle," as they are known, are the branch tasked with defending against chemical, biological, radiological, and nuclear weapons and were founded as the US Chemical Warfare Service (CWS) during World War I.

There were no native inhabitants on Johnston Island. Most of the people on the island were civilian employees of the company Holmes & Narver, which had provided the base operating services and support (BOS) since 1962. Holmes & Narver was responsible for all engineering, construction, and BOS at Johnston Atoll. BOS included work functions such as building, vehicle, and road maintenance; groundskeeping; security; fire protection; food and dispensary service; refuse collection; and operating and maintaining utility systems. Holmes & Narver provided the support at Johnston Atoll as part of an umbrella contract that included services at the DOE's Nevada Underground Test Site and its Pacific operations. The contract value for these services at Johnston Atoll was about $39 million in the 1980s. While Johnston Atoll never had any indigenous inhabitants, during its most active use, there were, on average, about 300 American military personnel and 1,000 civilian contractors present at any given time.

Chemical weapons were stockpiled on Johnston Atoll beginning in 1971, and 30 years later, in July 2001, the US Army Chemical Activity Pacific ended the use of the island and left. When Nick was on the island, it still had its complement of chemical weapon storage and other hazards. About two thousand tons of deadly sarin and VX nerve agents and blister agents were stored on the island in bunkers and managed by the Army Chemical Corp units who had special training in handling these hazards and had emergency responses in place should something go awry. As Nick understood it, the army kept cages of rabbits in the bunkers, and a dead rabbit was the first clue of a suspected leak.

A few months before, Nick sat in his office at 195 Sugg Road in the first-ring Buffalo suburb of Cheektowaga, New York, next to the Buffalo-Niagara Airport. He sat staring blankly at the papers on his desk, waiting for his coffee to cool. Because he was unable to sleep and tired because his sleep patterns had shifted so much over the past year from the constant travel and early start times, his coffee habit had increased. Whether it was spring, summer, or winter, they always started work early to avoid working through the hot sun in level B personnel protection in August or getting to the job site work trailer early in the winter months to start the kerosene salamander to thaw the tools, protective clothing, pumps, and blood in the near-zero temperatures. There is no way to describe the lasting memory of kerosene and coffee in the middle of the winter in the job site trailer on the Ashtabula, Ohio landfill with the cold wind running across from Lake Erie. But that is part of a different story.

It was early in 1984, and twenty-eight-year-old Nick was fresh off the dizzying previous year where he worked on well over 16 hazardous waste sites across the United States. He was reading through his somewhat daunting new assignment as he began to wrap his mind around the tasks ahead. He would be responsible for planning the trip, deciding what equipment was needed and procuring it, and getting himself and his crew safely to the destination on time.

They were going to Johnston Atoll for an almost two-month investigation of soil and crushed coral impacted with contaminants from the storage of Agent Orange from Vietnam. The equipment had to be ordered and brought on a barge to make it to the atoll, and the crew had to be brought in on military air-lift command (MAC) flights from Hickam Air Force Base in Honolulu, Hawaii, destined for the island on a certain date.

It was all Nick's job to pre-plan the trip, and he had just purchased six drills from Sears and about two dozen carbide hole cutter drill bits. These were 45mm, made of stainless steel alloy, rated for an extremely long lifetime, and specially designed to cut alloy and stainless steel materials. The brain trust helping Nick plan this adventure figured the drills would stand up to drilling into the crushed coral to collect the near surface samples planned for the job. He bought extra bits and plenty of extras of all the essential equipment because he only had a one-time window to get the stuff on a supply barge that only went to the island once a month. It had to be there, ready to go once they landed on the island. Not only did it have to last, but it had to work because there was no way to get anything else. The equipment list included all the various protective clothing—saran-coated Tyvek suits with attached hoods and boots, nitrile and butyl rubber gloves that would go over surgical gloves, air-powered air-purifying respirators with dust/organic vapor cartridges, other thick boots and the booties that went over them, and gloves, cloths, and rags.

Also, Nick had to ensure he had the medical monitoring equipment that would be used for heat stress monitoring and all the tubs, brushes, pails, and other decontamination gear and lots and lots of Gatorade in multiple flavors. Over the past year, Nick figured he had consumed enough of this fairly new product to last him a lifetime.

Nick first heard about Johnston Atoll a year or two before this assignment from his boss James B. Moore during his many "war stories" over a few pitchers of beer next to the office at the Holiday Inn bar. Never realizing he would find himself on Johnston Atoll, he sat somewhat in awe at the magical storytelling that big James B. Moore "Himself," as he was known, had perfected. James B. was one of Nick's mentors in the early days at Ecology & Environment.

He was an ex-Army Chemical Core guy who grew up in a military family and went to the Citadel for college. Founded in 1842, the Citadel is in Charleston, South Carolina. It has a noted reputation nationally for its Corps of Cadets, who live and study under a classical military system based on leadership and character development. Before E&E, James had worked at Roy F. Weston. Like E&E, Roy F. Weston was another early firm that provided a range of environmental consulting, design, and construction services in the new hazardous material and toxic waste environmental consulting arena.

James B. Moore "Himself" was a somewhat larger-than-life type of guy known for his gift of storytelling and presenting facts—even if some were a little stretched for emphasis. You know the type; he was tall with a full head of blond hair, square-jawed but a little overweight since his army days. He had the self-assured presence of an ex-military guy and a gift for gab. Nick had developed a friendship with James because they lived a few blocks from each other in North

Buffalo and had spent numerous hours together at training courses and hazardous waste sites. Having been stationed on the island while in the army, Jim would often mesmerize Nick and others as he regaled anyone in the group at the bar with his war stories—many of which were about his experiences on Johnston Atoll. Little did Nick know at the time that he, too, would have his Johnston Atoll experience, an experience that became even more real because of Jim's stories.

Nick forgets the exact story of how Jim found himself on the atoll, or perhaps he was never told. One of Jim's more memorable stories was about the time he had been left behind on the island as part of a skeleton crew during a hurricane to "take care of things" and how he was almost drowned in a jeep on a base road in the middle of the storm as waves were easily crashing over the concrete barrier wall that surrounded the island. Nick often thought back on that story when he was on the island; it had all the more meaning now that he was in that place in the middle of nowhere in the central Pacific—like a small bar of used soap floating in a large, vast bathtub.

The mission was to be manned by Nick and two others from the Buffalo office—John Duwaldt and Paul. The three had worked several nasty sites together, so the bosses figured they would be the ones to send. The twist was that a senior guy, Jake, from the Kansas City office, whom they did not know, was also going. After a couple of years in the field on sites across the country, Nick, John, and Paul were experienced hazardous waste workers and are fairly close friends. Theirs was a friendship honed through intense work in unknown hazardous environments and fostered by some hard partying together on their off nights.

As young and in shape as they were, they learned early on that anything more than a few beers the night before a 12-hour day in level B in a hot or even cold environment was more than stupid. These guys cared about supporting one another and about the job. They were part of the "group of geniuses" recruited by E&E in the early days of the 1980s, when environmental regulations were growing and new. This description of the collection of people that E&E headquarters had amassed was provided many years later at a company reunion when someone recalled the early glory days of the company.

As Nick was staring at the assignment, he momentarily wondered why this fourth crew member was coming from another office to be the lead. Perhaps this office had the military government connections that led to this project, or perhaps there was some other reason. The Kansas City guy was most likely ex-military because he acted that way the few times Nick had met him since his first meeting at the early training courses a couple of years ago. He came across to Nick as the type of guy who is all about chain of command and "being in charge"; someone who liked to bark orders and expected them to be followed.

Now that might work well for a young, inexperienced crew. But the guys from Buffalo had more than two years of working on some of the worst sites across the country and had developed a close friendship from being in the heat of some very hazardous situations together. This guy never picked up or adapted to the crew's working relationship, and Nick concluded early on that he never would. He was not a bad guy, and he certainly worked hard. But his air of superiority and the notion that the Buffalo crew was unruly and lazy was highly insulting to his Buffalo mates. These were experienced, trained, highly educated, and very hard-working guys. They did not take kindly to this military command nonsense. However, they managed to act professionally throughout, with Nick acting as the mediator because he had to room with this guy.

It put additional pressure on Nick, who shouldered it well, although there were many long walks at night to escape his room. One of Nick's favorite things while on the island was to find this quiet spot near the outside movie theater where he would lay on his back and gaze at the star formations so different in the central Pacific sky. Typically, on those nights, he would think back to his family and hometown: his friends, teammates, and lost love all those years ago. The view was just a wonderment as Nick lay in pitch black staring at the magnificent, unique celestial sky and pondering the past to escape the present.

The assignment was set up for Nick and Jake to go a week early, arrange for all the gear on the barge to be delivered to the job site, and set up a command post, shelter, and decontamination line. They needed to get the job going and have it ready so the crew could hit the ground running. Nick met Jake at the LA airport after arriving on separate connecting flights. They had a short layover before their flight to Hawaii, where they would spend a day and a half before they took a MAC flight to the island.

When they reached Hawaii, they spent a day sightseeing at military bases and the World War Two Arizona memorial at Jake's insistence before leaving early in the morning on the MAC flight from Hickam Airbase. Nick's main memory of this flight was entering a plane full of a sea of green and camouflage of military personnel dressed in fatigues for their stopover or assignment at Johnston Atoll. As they neared the island, the plane started to lower in altitude. Nick recalls his developing fear as the Pacific Ocean was getting closer and closer, and he felt the plane was going to land in the water. It was Nick's first experience landing on an island with nothing but the Pacific Ocean in view. Just as he was about to actually panic, he saw the tarmac as the wheels touched the ground.

Walking off the plane, they were immediately ushered into a structure and a room for instructions and then off to the nurse for blood tests. Now Nick was required to get blood tests once a year for his annual physical to measure any exposure and difference from the last year. Sometimes with an especially hazardous assignment, they had to get a blood test before the assignment and after to check to see if they were exposed on that specific job. In this case, it was a requirement of the military and being on the island.

Every time Nick got a blood test after his experience on the island, he would flash back to seeing the battle-axe military nurse shoving a large-diameter needle into his vein and being told that the military does not mess around with small needles. He had to get an exit blood test when they were done, and his arm was black and blue a couple of inches around his elbow for three weeks.

And now, just a few days later, Nick and his partner Jake were standing in the back of the pickup, looking over the cab at the two MPs in full chemical gear with weapons drawn across their chests, just staring at them. Despite repeated questioning from Nick and Jake when they first came to a stop, the two MPs stood motionless like the guards at Buckingham Palace, who are not allowed to speak or even flinch. "What in the world," Nick said barely audibly to Jake. There were no sirens—no warning that anything had happened.

Nick recalled the briefing they had gotten after landing on the island and immediately being ushered into a building. They got their briefing on all the island's nerve agents, mustard gas, and other hazards. They were told in no uncertain terms where they should never go. At this briefing, they were also issued and fit-tested to their mandatory military gas mask equipped with a carrying case and an Atropine kit, both of which they were required to have within reach 24 hours a day. The kit contained a hypodermic needle and two vials of Atropine. They were instructed that if they were exposed, they should jam the needles full of the counter agent into their thighs and "Kiss your ass goodbye!"

Nick and Jake looked at each other, sweating under the full tropical sun, and wondered what was happening. Suddenly a call came over the MP's walkie, and Nick and Jake could see a conversation going on followed by an abrupt "Yes, Sir." The MPs who received the orders looked up at his partner and declared, "all clear."

He turned to Nick and Jake and said, "You guys can go back to your work now." That was it; that was all he said. Not about to let this whole thing pass without some explanation, Nick said loudly, "What was that all about? What gives?" The MP calmly told Nick and Jake that a rabbit had died, and they thought they "had a leaker." "Since you guys were downwind, we could not let you come in contact with the rest of the base, and we had to observe you guys to make sure you did not have any symptoms." "Holy cow," Nick said to no one in particular. "Where were the sirens? Where was the warning? What about the vials of Atropine?" But he got no reply. The MPs just turned and moved off down base, leaving Nick and Jake to walk back. After that, Nick paid even more attention to his surroundings.

As the only shallow water and dry land area in 450,000 square miles of ocean, Johnston Atoll is an oasis for reef and bird life. This includes coral and coralline algae, about 300 species of reef fish, threatened green sea turtles, and seabirds such as the great frigatebird, red-footed booby, red-tailed tropicbird, sooty tern, and others. Johnston Atoll is also considered habitat for the endangered Hawaiian monk seal, a threatened coral species.

The atoll's military history began before World War II and continued until the military decommissioned the island in 2004. After Nick's time on the island, and by the mid-1990s, Johnston Atoll was the location of the Johnston Atoll Chemical Agent Disposal System (JACADS), which was used for destruction of the chemical agents that surrounded Nick when he was there.

Because of its military operations and the unique use of the island for radiological and biological testing as well as bulk storage of herbicide orange and chemical warfare agents, the island has had two Research Conservation and Recovery Act (RCRA) permits to store or treat hazardous waste. These included the aforementioned JACADS facility permit and the general Johnston Atoll permit. Because of the environmental impacts caused by its past, there are solid waste management units (SWNUs) and areas of concern (AOC) left on the island that require maintenance and monitoring under government corrective action. A SWMU is a physical space at which solid wastes have been placed at any time, either to purposefully store and manage solid or hazardous waste or by indifference. They include any area at a facility at which solid wastes have been routinely and systematically released.

The JACADS started and completed after Nick's time on the island and was the army's first chemical munitions disposal operation. It completed its mission on Johnston Atoll and ceased operation in 2000. The mustard and nerve agents were destroyed with high-temperature incineration as part of the JACADS operation. The metal weapons casings were thermally decontaminated and scrapped after the chemicals were removed and incinerated. The JACADS destroyed more than 400,000 rockets, bombs, projectiles, mortars, and mines. While Johnston Atoll is still under Air Force ownership and control, the facility, hazardous waste management units, and runway were closed in June 2004.

It was not all work and no play on the island for Nick and the crew. One thing about hazardous duty and the military was that they did everything they could to make the bizarrely hazardous conditions bearable. It's hard to explain the feeling of being on an island two miles long, a half-mile wide, and only seven feet above sea level at its highest point—surrounded by some of the most hazardous and deadly chemicals on earth. On the far upwind side of the island, there were volleyball and basketball courts, a very small nine-hole golf course with coral fairways and green grass landing areas, a full softball diamond with home plate facing the water on the far northeastern end so that even a monster shot would not make it through the steady 25-mile-an-hour wind over the fence to the water. During Nick's time on the island, he only saw one very, very large serviceman hit a ball that cleared the fence and went into the water. There was an Olympic-sized swimming pool, racquetball courts, and a small gym.

Food at the mess hall was just outstanding; it was served cafeteria style and was all you could eat. Wednesday night was prime rib night, and the steaks were served thicker and larger than Nick could eat. Nick particularly liked the variations of Kimchi served almost nightly and made differently by the various Asian cooks supplied by Holms & Narver. Saturday night was teriyaki steak night. If you chose to, instead of going to the mess hall you could go to the small beach area (labeled picnic area on the base map) along the north shore, which was set up for a cookout,. There, large stainless steel serving tables held large trays full of steak. Half were marinated in teriyaki, and half were plain. You picked your own steak and cooked it yourself, sitting under the moon and the magnificent starlit sky. It's hard to express how clear and different the sky was in the central Pacific. It offered a unique view of the constellations the equator offered.

They were each handed a copy of the Defense Nuclear Agency Johnston Atoll Base Guide when they arrived on the island, and Nick learned much about his new home from that guide. The guide had a chapter on the history of the island and its geography and chapters describing climate, general information, air conditioning, and the Red Cross. It also had paragraphs on the important stuff like banking, the barbershop, billets (temporary lodging), boats, the bowling alley, cameras, checks, clothing, and basketball. There was information about the store or PX, social clubs on the island, customs inspection, dental, obtaining your driver's license, the exchange, getting eyeglasses, fishing and golf, and a lot more randomly provided information. The content and sequencing of the guide must have made sense to the military, but it didn't to Nick.

Nick often wandered off on his own to get away from the pressure of the mission and his roommate. On one such occasion, Nick stood on the fishing pier on his Sunday off day gazing out across the relatively calm water in the lagoon inside the reef that rung the island. His buddies John and Paul must have been occupied, or he would have been with them, but he was there alone. His mind started to wander about life and how he wound up here—in the middle of the Pacific Ocean. In a US Fish and Wildlife article, he had just read that "Johnston Atoll is located some 450 nautical miles southwest of French Frigate Shoals, its closest neighbor in the Hawaiian Islands National Wildlife Refuge." He read, "Johnston and French Frigate Shoals may have played important roles as steppingstones for the migration of marine species between Hawaii and the Line Islands to the south." The article said, "Johnston Atoll is an oasis for reef and bird life and may be the most isolated atoll in the world."

An ironic smile fleetingly spread across Nick's face as he contemplated the totally incongruent reason for his being there. The fact that he was required to carry his Army full face mask, the hypodermic, and vials of Atropine with him wherever he went on the island should there be a leak was juxtaposed in his mind against the wonderful wildlife and habitat he had just read about as the current scene played out before his eyes. He did not tell the Army Chemical Corp MP during that encounter a few weeks before how silly it was since any exposure would require more than this ill-fitting military Army mask to protect him.

The clothing they had shipped via barge was much more protective than the Vietnam-era stuff the military was wearing, and the Army briefing and procedures would be fruitless if such a leak occurred. He guessed that was what the "kiss your ass goodbye" was all about. This was the problem of knowing too much when surrounded by an authority that really knew little. It was much like Nick's experience with both the government and the media that sold their idiotic programs. "Oh well, I'm thinking too much again," Nick told himself. "Just get back to the beauty and stop letting your mind try to fix all the world's problems." Nick often needed escapes like this to cope with a world he increasingly realized he understood more about than most. It was the politicians that affected him the most and their hypocritical drive toward power. That people were so easily led by shyster politicians was more than an annoyance to him. As he gazed out at the serf pounding the outer reef as huge waves rolled across it, his mind welcomely shifted back to nature.

The marine environment around Johnston Atoll consisted of a shallow coral reef platform encompassing approximately 50 square miles. Nick had further read that "Johnston Atoll is unlike most coral reef atolls: the protective ridge of coral reef extends only along the northwest side due to its underlying platform subsiding and tilting southeast. Most of the reef lies outside the lagoon, extending approximately 11 miles east-southeast and 5 miles south of Johnston Atoll." That was why Nick's only knowledge of how huge the waves were he saw really came through high-powered spy glasses.

Nick had bought aerial photos of the island, clearly showing it was nothing like its original land mass and was more man-made now. It looked like a giant aircraft carrier with a runway running its length. He read that Johnston Atoll had been enlarged and shaped twice by dredge-and-fill operations, the first in 1949–1950 and the second in 1963–1964, resulting in one square mile from its original dimensions of 0.07 square miles.

Nick loved nature and biology, a fact that drove his older brother nuts when they were young because Nick insisted on watching every nature show that came on their sole TV. The island was visited annually by the US Fish and Wildlife Service, and when Nick was on the island, they dropped off several seals. Hawaiian monk seals have been listed as endangered since 1976. The Hawaiian monk seal is native to the Hawaiian Islands archipelago and Johnston Atoll. NOAA Fisheries was working to protect this species in many ways, with the goal of increasing its population.

Nick read that they observed that a significant cause of female and juvenile monk seal mortality during that time was aggression from multiple male seals, especially at the Kamole and Kapou Islands and the French Frigate Shoals (Lalo) and Kure Atoll (Hōlanikū). One day, Nick noticed a NOAA research ship in the big boat harbor of the island. He was told by their scientists that NOAA Fisheries found that removing specific aggressive males prevented the many killing of females and pups. Nick took photos of one such male seal that NOAA dropped off as it lay on the beach near their work area.

As he gazed out across the reef, purposefully forgetting the irony, he became lost in the extent of the natural beauty teaming with wildlife he had never had the privilege to see before. Green turtles nested and basked on the south shore beaches, which they also used as a feeding ground. At certain times, the atoll housed the highest concentrations of green sea turtles in the Pacific. The Fish and Wildlife article about the Island that Nick devoured informed him that most of the marine mammals are visitors just outside the atoll's reef, with few inside the lagoon waters. This included bottlenose and spinner dolphins, Cuvier's beaked whales, Hawaiian monk seals, and humpback whales.

Nick read that "approximately 300 species of fish have been recorded in the near shore waters and reefs of Johnston Atoll." And there were many different interesting, beautiful birds, including the black and the deep chocolate brown noddy with their stereotyped head-nodding courtship displays found mostly on Sand, North, and East Islands, with many pairs nesting in trees on the main part of Johnston Island. Other birds included the migratory bristle-thighed curlew, the small sooty-brown bulwer's petrel, Christmas shearwaters, gray-backed terns, the great frigatebird—the largest seabird on Johnston Atoll, which is the most efficient of soarers as it glides on the wind or thermal updrafts, stealing its catch or unattended eggs.

Other birds included the masked booby, the migratory Pacific golden plover that breeds in western Alaska and Siberia and winters on the islands across the Pacific Ocean, the bright red-footed booby with its light blue beak along with the red-tailed tropicbirds, ruddy turnstone that often forage by turning over stones and other objects, the wandering tattler, and the wedge-tailed shearwater.

Nick's favorite bird on the island was the white tern. These snow-white birds had jet-black bills. Nick first noticed these birds when he was walking across the off-limits area of the former herbicide orange drum storage area. He was fully dressed in level C—complete with a pure white saran-coated hooded Tyvek suit and a full-faced positive pressure mask with its black hose leading from the front of the mask to the powered four-cartridge battery pack worn on his hip. It was Nick's first or second day in the "hot zone." He and others were laying out the grid pattern using 200-foot measuring tapes to mark out the area for the future sampling. As Nick walked across this former agent orange storage area at the furthest downwind northwest corner of the island, two small pure white birds with black beaks hovered just above his head, dipping down to peer into Nick's eyes. Nick wondered if these birds had somehow mistaken him for a long-lost father in his all-white suit and black-hose "beak."

He heard later that these birds have a habit of fluttering curiously over visitors to assure that their presence will be noticed. The entire body of this tern is white with a black eye ring, which creates the appearance of large eyes. The thick bill is mostly black with blue at the base. Its legs and feet are slate blue, with yellow to white webs. The tail is shallowly notched. They breed throughout tropical and subtropical Pacific, Indian, and South Atlantic Oceans.

Leaning against the railing along the pier, Nick was engaged in looking at all the tropical reef fish and especially the coral snake all around the remains of some fish that the last sports fisherman caught and gutted. Some military crew—mostly the higher ranks—often took the sports fishing boat out to go reef diving or sports fishing. Nick never went because he only had Sundays off. He was watching the black and white striped coral reef snake when he saw the black tip and the undulating swimming motion of the black-tip reef shark. As it came toward the remains of the fish guts, the coral fish and even the snake scattered quickly away. Just as he was focused on the shark, he was taken out of his gaze by a corporal who had snuck up beside him and proceeded to hum the theme song for the movie Jaws; a chill ran down Nick's back. Black and white-tipped reef sharks are two of the three most common sharks inhabiting the reefs of the Indo-Pacific, with gray reef sharks being the other.

On one occasion, Nick went with Paul and John to the island's southwest end, where the garbage chute was, to watch the large sharks. When the garbage was emptied, the sharks came rushing in to feed. You never saw these large sharks on other parts of the island because the garbage chute was located purposefully on the part of the island where the reef was minimal. This allowed the garbage to wash out to sea, but it also allowed these large-bodied sharks a chance at a free meal—one they had become accustomed to, almost like feeding time at a zoo. It was quite a memorable sight watching these large gray sharks in a feeding frenzy on the leftovers from the mess hall.

The unofficial flag of Johnston Atoll is a double bird holding four stars. The birds represent the Air Force and the Fish and Wildlife Service, while the four stars denote the atoll's islands. The white on the flag is for coral, and the aquamarine is for the surrounding ocean. By 1964, dredge and fill operations had increased the size of Johnston Atoll to 596 acres from its original 46 acres and also increased Sand Island from 10 to 22 acres, adding the two new islands, North (25 acres) and East (18 acres). The reef crest on the northwest portion of the atoll provides a shallow lagoon with depths ranging from 9.8–32.8 feet.

The climate is tropical but generally dry. Northeast trade winds are consistent, making the southwestern end the downwind side of the island, and a 15–25-mile-per-hour wind blew steadily in that direction about 99% of the time. Little seasonal temperature variation occurred, and it appeared to Nick to be a pleasant 72 degrees most of the time, but it felt hotter under the tropical sun. The island had mostly low-growing vegetation, with a few palm trees on the mostly flat, desolate terrain.

While on the island, they worked 10 hours a day, six days a week, with Sundays off. At night in their free time, they either watched the newly installed satellite TV and HBO or went to the nightly movie at the outdoor movie theater, which consisted of a series of rows of stadium seats and an outdoor Drive-In movie screen. Sometimes they would go get drinks at the club. The Waikiki and Tiki clubs were open for off-duty enjoyment, and Nick and crew had just missed the USO show that only visits once a month. There were various other clubs on the island that required membership to get into.

Most were small huts run by the different ethnic populations, mostly Asian—Chinese, Filipino, Korean, and Burmese—that made up the civilian workers. Nick, John, and Paul usually went to the aforementioned Waikiki or Tiki clubs that were open to the service population. But they also sometimes went to the JI Social Club. By chance, when they arrived on the island, it was the beginning of a volleyball league. One particular team, the JI Social Club, was short three players. Nick, John, and Paul (who had played competitive volleyball) joined the team. Being natural athletes, they helped the JI Social Club win the league with Paul's instruction. The boys were made lifetime members of the JI Social club, complete with T-shirts to commemorate them.

Nick had also befriended the Holmes & Narver backhoe operator, who was sent to the job site to dig a series of test pits across the Agent Orange area in between the grid sample locations. Nick and the boys were often neck-deep in these trenches sampling the coral at depths down to five feet or more. As they were sampling or during break time, Nick would chat with the backhoe operator about all sorts of things, in typical Nick style, and they kind of hit it off. The Hawaiian backhoe operator's name was Akoni, which means "invaluable," but for some reason, he went by the name Sam. Now he sometimes slid into slang Hawaiian called Pidgen, especially as he became more comfortable with Nick. Nick was lost for a while but started to catch on as the weeks went by because the slang retains a higher degree of mutually intelligible portions of words, so once Nick got the flavor, he could follow—somewhat.

Hawaiian Pidgin is an English-based creole slang that originally evolved during the times of the early sugar industry, which began in the mid-1850s on Maui. The language has parts of Chinese, Japanese, and Portuguese from the first wave of immigrant workers and blends of other languages. It continued to evolve as waves of Okinawans, Puerto Ricans, Koreans, and Filipinos arrived around 1900. Sam had been on the island for at least ten years.

He told Nick, "I have eight kids. Seems like every time I go back to Hawaii, I get my wife pregnant." He went on: "I go back to Hawaii once or twice a year or when a hurricane comes and they evacuate the island." Near the end of the project, Akoni (Sam) invited Nick to come to his Pacific Islander club on a special invite. Nick could not wait to check out one of these other clubs and spend some time with Sam outside of work.

Nick took the bike he was issued and rode over from his barracks to the area where the clubs were located toward the northern side. When Nick entered the club, he was introduced to a couple of guys sitting around a table in the middle of the club. Then Sam introduced Nick to his best friend. "This is my friend Keola, which means 'live one,' but everyone calls him Burt."

After the formalities, and as Sam shoved an ice-cold beer toward Nick, Keola (or Burt) said, "So, you catch any fish when you were fishing earlier after work?" Sam replied, "Oh yeah. Choke, brah." (Choke means "lots of.") Sam continued, "I just wish Nick here didn't keep me so late, or I would have caught more." Then Sam explained to everyone there what Nick was doing on the island and how they met. Most of this Nick understood because Sam had slipped back into slang while talking with his clubmates. Burt, seeing that Sam slipped into slang, replied in an obviously good-natured way, "Stop that complaining. You so irrahz [irritating, annoying]. C'mon, gimme some more, brah. You so chang [cheap, miserly]."

The club was a two-room structure with a lanai and tables outside. One room was a fully equipped kitchen, and the other was a living room with a bar and TV. Nick spent a couple of hours drinking beer and talking while Sam and his buddy worked on their stained-glass projects from a course they had just taken. Apparently, every month or so, they took some new course given on the island. Nick never understood how someone could spend that many years on this small island of mostly men with military backgrounds or escaping someplace or something.

Most of the people Nick met on the island were a little off, and everyone spoke in what Nick termed "military lingo": "who ordered this weather" or "balls to the wall" or "boots on the ground"; "bought the farm"; "caught a lot of flak"; "roger that"; "no man's land"—as in "you guys are working in no man's land," meaning the off-limits Agent Orange area. There was this one guy who seemed to crop up from time to time as the boys spent the weeks on the island. Nick and the guys decided he was just a little squirrely, and it was safer to be a little wary around him. He always spoke in this kind of military lingo and would typically greet you by saying, "Hey, who ordered this rain?" If it weren't raining, he'd ask who ordered whatever was occurring at the moment you ran into him.

The one thing that Nick remembered the guy saying, maybe over a few beers at the Tiki Club, was that he wanted to be as far away as possible from mainland civilization, and he figured Johnston Atoll was the perfect place. They never did find out what the guy was running from, but Nick always had the feeling it wasn't good.

Nick started every morning reading the island's "newspaper," The Breeze, which amounted to a single page printed front and back on yellow paper. Nick had saved one random paper to bring back home The potential exposure Nick and the crew got during his time on the island is unknown. They had spent most of their time in the Agent Orange area neck deep at times in the crushed coral soil that had been saturated with herbicide and probably had remnants from the nuclear testing fallout and biological agents. Over the years, Agent Orange and chemical weapon leaks occurred in the weapon storage area. Multiple studies of the

Johnston Atoll environment and ecology have been conducted, and the atoll is likely the most studied island in the Pacific. Some polychlorinated biphenyls (PCB) contamination in the lagoon was traced to Coast Guard disposal practices of PCB-laden electrical transformers.

About 45,000 tons of soil contaminated with radioactive isotopes was collected and placed into a fenced area covering 24 acres on the north side of the island. The area was known as the Radiological Control Area but dubbed "The Pluto Yard" because of its heavy contamination with highly radioactive Plutonium. Remediation included a plutonium "mining" operation called the Johnston Atoll Plutonium Contaminated Soil Cleanup Project. The collected radioactive soil and other debris were buried in a landfill created within the former LE-1 area from June 2002 through November 11, 2002—long after Nick's crews' time on the island. Remediation at the Radiation Control Area included the construction of a 61-centimeter-thick cap of coral sealing the landfill. Permanent markers were placed at each corner of the landfill to identify the landfill area.

In 1995, the U.S. government declassified a set of military documents that outlined the real story behind a series of failed high-altitude nuclear tests and atmospheric air drops during 1962 on this tiny atoll in the middle of the Pacific. This stoic silence on the part of these brave soldiers in service of their country reveals the absolute secrecy surrounding these tests, which was played out on this forgotten atoll in the mid-Pacific. The men who participated in the 1962 tests suffered an 85% casualty rate from various radiogenic diseases. As previously mentioned, one pad disaster spit Plutonium over most of the western part of the island, including the first 300 feet of the runway, the launch area, the parking area, the swimming pool, the cafeteria, and the latrine. Within 25 years, most of the men trapped on the island during the testing would be dead from diseases believed to be a direct result of radiation exposure from the accident.

By May 2005, almost all of Johnston Atoll's infrastructure had been removed, and all personnel had left the atoll.

Chapter 2
HE WANTED TO BE GREAT

Still just kids, really, they started on one of their many walks in quiet, taking in the beauty of the forest path. Nick finally admitted his feelings to her, ending the silence between them. Neither of them were talkers, so walking along together in the forest without words was not at all uncommon. It happened somewhere along the path through the woods that led from the back of the lake past the giant boulders and rock formations that the locals called Cave City. Nick glanced over at her as they walked, formalizing into words what was evolving in his head.

As he spoke, his hands gestured in a way that emphasized his words, which was not common for Nick. And so, she knew he was passionate about what he was saying. He told her he wanted to accomplish great things, make a difference, help the world in some manner or act—be a great man! This is something he could only tell his girlfriend or maybe his best friend, Scott, who knew him well enough to accept these words for what they really meant. That was it; that was all he said, and her only response was to slip her hand in his, again walking among the tall trees in silence. There was no need to question him about his statement. She knew him and knew what he meant. She knew that if he felt like explaining it further, he would probably somewhere farther down the path as they were entering some older growth maple and oak forest that had probably not been cleared since colonial times.

It all started on the side of Joe's Delicatessen, at the foot of Varian Road where it intersects with Oregon Road just outside Peekskill, New York. This is where he did a lot of his thinking as a young man while drinking an ice-cold orange Nehi soda that he collected from the cooler just inside the front door of the deli. His desire to be great had nothing to do with being idolized or wanting fame. His need to be great was about a need to have some positive effect on humankind. Without ego, there is no greatness. Nick certainly had an ego, but it was a mature ego that should never be confused with being egotistical or self-aggrandizing. No; Nick was much too self-reflective for that. He beat himself up far too regularly, because he understood his flaws too well. The ego was there even at his young age, perhaps stronger because life had not yet battered him around. Without ego, there would be no accomplishment. Even so, as his life trudged on, he lost track of this desire although the years never diminished the haunting lost first love.

His mother always told his brothers and him that they could be anything they wanted as long as they worked hard. This lesson, along with his father's of doing a job right and never halfway, were engrained in his soul. The progressive socialism that sapped people's self-worth had not yet made its full-throated attack in America. Having a great stable family and upbringing is so crucial to reaching potential; it's not essential but is certainly a major advantage. It's something he would learn firsthand later in life when he befriended several people in the inner city of Buffalo. It hurt his soul seeing innocent kids with such a disadvantage in life as political policies had ravaged the poor, which they falsely claimed to be helping.

When he was older, he never really understood the desire for greatness in either a conscious or subconscious way, and he never knew when he achieved it—or if he ever did. Was it his egotistical guilt of being afraid to show superiority in a braggadocious way that he took such umbrage to accepting its implications from others? He never liked the preening or swagger of some in their accomplishments.

Knowing it lied within him, he routinely belittled himself. Suffice it to say that he did not take compliments easily, because he never gave himself the leave to accept them. Craving platitudes, he withdrew when confronted by them. Was all this part of the sliver of his ancestry with a race more guilt-driven than any other people on the planet? Certainly, he was not this smart, as others suggested he was, because he knew himself so well. He knew his limitations, and he tried to hide those that were overtly apparent to him. He used this to keep himself in check, and when he failed to check his ego those instances became the more lasting memories that he took with him. He was a very conflicted soul at times, but in the end—perhaps much later than it should have—it didn't hamper him greatly. His confident personality ultimately made sure of that.

At times, he struggled with simple easy things that appeared so effortless for others. He was unable to spell because his brain often mixed-up letters and words that made so much sense only to be so clearly wrong when he re-read the editor's red notes on some technical paper he was writing. Grasping so hard for the right words for the concepts his brain perceived so clearly often slowed his progress as the brilliance faded in his lack of coherent expression. Eventually he would find that brilliance, but often it meant he had to take extra time. Often, he would race to get down on paper or on the computer screen the brilliance that he knew was fleeting in his mind. Sometimes he did not make it, and whatever vividness his mind perceived was lost in his inability to express it. It became muddled; clear to him in only a fugacious way. It was often just so annoying to him that with time and age he became better at not losing the transient brilliance when it appeared, sometimes out of nowhere, in his mind and he would race to jot some of his thoughts on paper before they became mangled again in his mind.

He would of course recognize it, and it came often when he was not trying. Sure, he developed coping mechanisms that deceived most, but all this made clear to him his weaknesses and convinced him of how much he lacked. He certainly was not a savant because his mental disabilities were nowhere close to significant. But, especially later in life, he recognized certain abilities that were not greatly above but were certainly in excess of average.

The thoughts at times were so overwhelmingly resplendent and clear, even he began to give himself credit before he would reproach himself for being pretentious. He had a sense that his cleverness and astuteness was not normal, but he didn't develop it. So much was left undeveloped as a young man. When he played football, he'd see the play developing almost before it actually happened. He assumed everyone did also, not realizing until many years later when he coached that most others didn't possess the gift he had. His flaws more than kept him humble, making him feel nakedly inadequate, especially during the fleeting moments when the slippery thoughts vanished before he could accurately describe them.

He had trouble with basic addition and subtraction, even though the mathematical concepts were crystal clear to him while he used his fingers to get the right answer. Nick had developed coping methods to overcome his many shortcomings without even understanding them. In today's lexicon, they would be viewed as learning disabilities to the well-meaning but intellectual lacking teachers. It was all very funny to Nick that people, mostly those with weak minds, just assumed a lack of intelligence if certain people could not express themselves clearly in spoken or written words because they lacked the benefit of a good formal education, speaking and writing in poor white trash or ghetto language because that was the language of their friends. This perhaps is why Nick so easily moved in the different worlds he found himself: the inner city, the bayou of

Louisiana, the poor Appalachian counties of New York, on a military base, or at an elitist snob-filled Yale cocktail party. He had to catch himself sometimes when he did his master's degree in the public health department, a part of Yale Medical School, because he would almost start to say something in the slang he had learned working on a garbage truck.

Or, it was just as inappropriate when he would slip and use "Yale" words when he was with his buddies at his hometown bar, Mutt & Jeffs, or in the garbage truck yard talking to the other workers. It was all rather funny looking back when he sometimes failed, watching the expression on the "elite" Yale students and faculty when some garbage truck slang would fall out of his mouth, or he would revert to the phraseology of his youth.

One such memorable instance was when he was asked by his fellow students at their Ivy League professor's party what he did on his summer vacation. He purposefully decided to tell them, "riding on the back of a garbage truck with the hopper juice [liquid/water that accumulated from the garbage in the truck hopper] and maggots and rats." Then he wondered why he thought that was a good idea. Maybe it was just an inappropriate rebellion from these people who obviously had no other life lessons in other parts of their America. Their contrived interest rapidly fled as his story registered in their minds, and they quickly went back to their stories of Mediterranean cruises or trips to Europe.

All that drove him to be better, to overcome, to excel, to be the better man even though he knew he never would be the best. He was often awestruck at the clarity of others' brilliance, including his brothers, who seemed so much brighter than him. It was both a motivating factor and an unobtainable goal that he knew he would never obtain even though others thought he had.

Did you ever want to be great? My quest was to lay out more of Nicks life to gauge whether he obtained it or if it was a noble cause to begin with. For Nick, he had no choice; greatness was both inherently sought for and denied and never, ever flaunted. His drive to be great, a consequence of his heredity and upbringing, would bend along throughout his life consciously and subconsciously as he worked toward it. He was competitive in sport and study and business and was seldom happy even in victory. His motivation even in his lowest times were his mother's words: "You can be whatever you want if you work hard enough." I have often wondered when clarity of thought represents genius or deception. Were Nick's thoughts sagacious or merely prejudicious? My conclusion is that no matter how resplendent or how pure their conception, it remains for the neutrality of others to judge their genius, and that is what I set out to do.

I see racism today like I saw on the films from the 40s, 50s and 60s. It's coming from so called progressives, and it's as bigoted and racist and acceptably out in the open—perhaps more so—than any that existed in the past. All these racists have done is rearrange the deck chairs from their racist past. It's preached from the political pulpit using words like diversity and white guilt and white privilege.

It's put forth in WBE/MBE requirements and in many other ways, such as Critical Race Theory. I walked out of a restaurant/club Friday night and saw a white male being taunted by a group of "men of color." As I walked out the door, I heard one of the men state, "We need to teach white people a lesson," as he eyed me. He seemed to assume I'd find this acceptable. I took a few steps out the door and around the corner thinking I would approach the white man escaping this overt racism when instead I decided to go back and confront the group. Their response was to reply to me with more bigoted and racial nonsense before the bouncer, who was part of the endearing group, decided to deescalate the growing anger.

This openly aggressive and now acceptable bigotry and racism betrays the progress that preceded it. It's being used as a wedge issue by the same people who precipitated the ugly bigotry and racism of the decades past. And so, the division becomes magnified, and an evil person drives over 100 miles to Buffalo, New York to shoot black people as if we have learned nothing while the politicians use it to cement the division.

My name is Jedidiah—a biblical name that traces back to an ancestor long ago. I just completed a detailed ancestral record for Bailey McCarthy and his best friend, Croft Evans, and was asked to focus on the life of Bailey's ancestor Nick McCarthy to fill in the story from where I had left off in his youth. I chose to focus looking in depth through a window in the house of his life associated with his profession as one of the early hazardous waste workers. This was one of the most influential aspects of his life, and it truly defined Nick.

I am a master historian and familiar with the geographic and historical record of places. I have extensive knowledge of genealogy and DNA database research. Even though my main purpose of the previous assignment was to create an extensive ancestral, genealogical, and historical record of Baily's makeup through a study of specific parts of the familial tradition, I had left off at Bailey's ancestor Nick's post college graduation. This was my assignment now and what I found was sheer serendipity because the occurrence, development, and events of Nick's life happened seemingly by chance—as is the case in most lives—or perhaps it was just ironic cruelty. How else would you describe the outcome of life mirroring academic study?

The friends and acquaintances he met living his life's purpose foretold their collective outcome. You can't get out of this world alive, which was never a consideration during Nick and his friends' youth. Was the outcome of their lives brought on by the things they later studied? Did these put them in the position for the pathways that most likely ended their lives, or was there much more to it?

My own genealogy and DNA confirmed that I was almost 80% African, including Nigerian, West African, and Kenyan, with some connections to the Maasai tribe—a proud people known for their fighting prowess and warriors. But I also have 20% British and Irish ancestry. During the past assignment I became very focused as the past played out before me. The boys' spirits and ancestral history became real—as if I were right there in real time with my ancestors who lived at that time. But this assignment required a much deeper look at the events of Nick McCarthy's life to go along with the ancestral heritage.

Since Nick was born into and lived his life in such a transformative era, understanding what motivated his purpose and eventual death is uniquely interesting to me as a historian. The times of the conservative 1950s, when he was born, led to his adolescent years spent in the tumultuous 1960s and 1970s replete in its counterculture and divisiveness. This then led through the nondescript 1980s and 1990s, which seemed in a way to be a buffer or, better, a reprieved calmness before the full-on looniness of the 2000s.

I again became quickly engrossed in my work, becoming one with the detail, and found myself in Nick's time almost like an unseen observer. Because I believed in the concept of rugged individualism, this assignment would allow me to study the life of a rugged individual as he transected through and transcended the era of progressive lunacy. How he survived the insanity of progressive education and was able to still maintain the ability of logical thought progression was of keen interest to me. It became clear that the times in which he lived created added roadblocks to reaching greatness because clear thinking was muddled in so many ways, almost purposefully by educators and leaders as they contorted, conflated, confused, and outright made-up "facts," which they put forth in righteous determination to cancel out all other competing more logical thought.

In his later years, you could sense that the mind-numbing stupidity around him left him weary. How did he become who he was despite his education at excessively progressive institutions? Perhaps it was because he was engrossed in the "hard" science and math courses that were based on logic and the scientific method well before the 2000s, when even these subjects were inflicted with inane ideology and warped into nonlogical entities. Undoubtedly, his heredity and upbringing played a chief role.

This led me to take a deep dive into Nick's genetics. From his recent ancestral record, I knew he had grandmothers who were first or second generation from Ireland and a grandfather whose great-grandfather emigrated from England. What was a little murky was where his other grandfather's father came from. Family rumors had him emigrating from Latvia to Germany, where he married a woman from Germany before immigrating to the United States. This is where the Ashkenazi part of his heritage obviously came from. Nick had the interesting "American mutt" recent ancestral background of Irish Catholic, English Protestant, and European Jew.

Nick's ancient ancestral background was initiated like all of us from the African content with a so-called "Eve" originating from the region of Kenya and the eventual maternal "Haplogroup I," which originated in Eurasia (mostly Iran) nearly 30,000 years ago. A haplogroup is a collection of common inherited genes that trace male or female ancestor lines back to an original parent. One of the first haplogroups to move into Europe, "Haplogroup I" appears on average in less than 2% of Europeans and under 1% of Near Easterners today. Nick's paternal, or male, ancestry "Haplogroup J2" traces back to an "Adam" from somewhere in the Cameroon region, with his ancestors from the "J2" group originating from the region of the Fertile Crescent in Syria.

My own direct recent ancestors lived in Cortlandt; the town Nick was originally from. For my last assignment, the head of the department said, "It was thought as the icing on the cake of a long illustrious career for you to focus on a case from the area of your ancestors." What I did not know, but would come to find out, was that my own ancestral history weaved through that of Nick's here and there throughout his ancestral timeline. Totally unexpectedly, the connection jumped out during my last assignment, when I studied the ancestral makeup of Nick McCarthy.

My deep-dive analysis into Nick's genetics also provided information on his more recent ancestry, or direct DNA, which showed he was almost 98% European with some outlier admixtures, such as Japanese and Punjabi in his advanced timeline within five generations. The advanced timeline is designed to show statistical outlier traits that are inherited from an ancestor in a given generation, using the 642, 824 unique genetic markers and generational decay algorithms that read chromosome segments and sizes to produce the findings.

Genetic admixture occurs when individuals from two or more genetically distinguishable groups have children together. This might happen when individuals from one part of the world settle into a new geographic region already inhabited by other people, e.g., due to invasions or large-scale migrations. Admixture analysis, more properly known as biogeographical ancestry analysis, was one of my focuses in school long ago. It is a method of inferring someone's geographical origins based on an analysis of that person's genetic ancestry. Apparently, along the way within the past five generations, some ancestor of Nick carried those outlier gene segments.

A large portion of Nick's recent ancestry based on his DNA was from the British Isles/United Kingdom (UK), which represents the entire region of the UK, including Ireland and Scotland. This of course matches up with his known recent heritage of two grandmothers from Irish ancestry and a grandfather from English ancestry. However, the genetic heritage of the people from the UK is so diverse because they came from Anglo Saxon, Norman (French), and Celtic lands, to name a few. Celtic cultures were the first to dominate the region known as Germany today. While many people associate Celtic cultures with Ireland and Scotland, Celtic cultures are believed to have originated in Austria and Germany and spread outward from there. As far back as 2,500 years ago, Celtic cultures dominated Germany and many other regions of Western Europe. And this also explains the other large portion of Nick's ancestral findings: German.

The murky knowledge of his remaining great grandfather's heritage, believed to be from Latvian Jewish ancestry, possibly explains the almost 20% Ashkenazi Jewish findings. It has been reported that "In the late Middle Ages, due to religious persecution, the majority of the Ashkenazi population shifted steadily eastward, moving out of the Holy Roman Empire and the Iberian Peninsula into what is today Belarus, Estonia, Latvia, Lithuania, Moldova, Poland, Russia, Slovakia, Ukraine and Germany. During the Jewish Diaspora—or migration of Jewish people from the Middle East to other parts of the world—the vast majority of Jewish individuals married and raised families within their faith. Ashkenazi Jews are a Jewish ethnic group who have their earliest ancestors from the indigenous tribes of Israel.

Studies have found that 50–80% of the Ashkenazim DNA from the paternal lineage originated in the Near East. But the interesting thing about Ashkenazi Jews is that the maternal line comes from Europe. It is apparent then that while the men migrated from the middle east, they married European women and over many generations their recent DNA appears mainly European.

Understanding Nick's ancestral heritage was an important aspect; it was a large piece of the puzzle to understanding who Nick was. I used this base to weave the story of Nick while life had its inevitable effects.

Chapter 3

I DON'T LIKE WATCHING YOU DIE BECAUSE I KNOW I CANNOT BRING YOU BACK

Prior to my leaving for Western New York, my friends had a going away party for me at our favorite hang-out. The bar had cordoned off a back area of the establishment for the party. As older folks, our parties usually started in the early evening and never lasted much past 12 midnight. They often ended some hours before. I would have had the party at the former Mutt & Jeff's bar of Nick's youth, but that bar on Oregon Road along with the drive-in movie theater across the street had long ago been demolished and replaced by condominiums. The party was somewhat fortuitous, for as I was contemplating my move to Buffalo to discover Nick's journey and life in Western New York, I found that Nick's friends had had a going away party for him. His party started at the house of his childhood friend Eric Schrull and later wound up at Mutt & Jeff's, where things eventually always ended up back then.

I arrived a little early before the party was officially to start to meet my friend Barry, who was opening up a bottle of my favorite bourbon: Michter's 10 Year Kentucky Straight Bourbon. "This is 94.4 proof limited edition bourbon, Jedidiah," Barry said in kind of a half wise assed way, using my full name instead of his usual "Jed" as he filled two glasses. "It has been rated best American Bourbon and is the favorite for most of the premium and elite bourbon connoisseurs." "What do I know from connoisseurs? I just found that I really liked the stuff," I said to Barry as he handed me one of the glasses of Bourbon with just the right amount of ice.

The smirk on Barry's face expanded even more as he said something about my expensive taste. He suddenly got serious, raised his glass, and said, "May the saddest day of your future be no worse than the happiest day of your past." "I'm going to miss you, my friend." He then asked me how my research was going followed by "Why in hell do you need to move to Buffalo of all places?"

It was a blessing to have such good friends, and I hoped I would meet some great ones in Buffalo. I proceeded to tell Barry that I was not sure how long I would be in Buffalo but that it would likely be no less than two years and maybe longer. "Barry, I am kind of following in the footsteps of Nick's journey when he left Peekskill after he finished graduate school and moved to Buffalo at age twenty-five to pursue his career. My research showed that they threw Nick one hell of a going away party when he left. Tell you what, Barry: I certainly have no plans to stay as long as Nick did in Western NY, and I'll be back to my Hudson Highlands home drinking bourbon and going on great kayak trips with you soon."

Just then I recalled an event when Nick was in his early 20s and recounted it to Barry. One of the members of his childhood gang died. He was driving too fast on his motorcycle and failed to make the severe curve on Oregon Road when it makes that sharp turn onto Oscawana Lake Road where Westchester County and Putnam County meet just before Putnam Valley, New York. He was on his way home to his apartment on Lake Peekskill. Lake Peekskill is a small manmade lake that was originally called Lower Cranberry Pond. A small unnamed tributary of Peekskill Hollow Creek, the creek Nick fished for trout in during his youth, was dammed on its southwest end in the 1920s creating a lake as a destination recreational area for people from New York City. Summer cottages adjacent to the new lake were sold to city residents as an escape an hour north to "upstate." Over time, the cottages developed into homes for year-round residents, and it became the hamlet of Lake Peekskill.

At the wake, looking down at his friend in his casket—the first time he had this experience—Nick wanted to just shake Tommy and wake him up. It made a lasting impression and later he thought how devastating it must be to the minds and souls of soldiers watching their best friends killed next to them. He had such respect for the men and women who found themselves in war.

I woke up early the next day ready to go. I made sure not to drink too much the night before—a lesson learned repeatedly, or so I thought, in my youth. I planned on a big day in the morning, one I wanted to do alone and not hung over. After completing a deeper dive into Nick's genetics, my plan was to start by gathering some basic facts about Nick's life after childhood. For this and the work that proceeded it, I did most of my research from my base in Cortlandt, New York. Within a week, however, I planned to move to Buffalo, New York, to get a sense of where Nick would spend most of his career and life. Before starting the process, I decided to take a kayak trip in the place of Nick's youth and college years. I decided to kayak into Annsville Creek in the lower Hudson near Peekskill, New

York. The Peekskill area was part of the life of Nick's youth. Today, the Annsville Creek Preserve contains a walking trail near the majestic Hudson River and its tributary, Annsville Creek, and includes a fishing pier, historical plaques, and a gathering circle. More interesting to me was the launch that is part of Annsville Creek Paddlesport Center. Some Kayak tours are run out of this spot along with a small outfitting and touring concession. The entry point is from a large interlocking plastic floating pier that appears to be designed more for jet-skis. Although not nearly as kayak friendly as launches that I later discovered in and around Buffalo, designed specifically for kayaks with rollers and handrails, this was better than trying to exit at the river's edge.

Prior to taking the trip, I reacquainted myself with the history of Peekskill and Annsville Creek—which was in my nature. The quick history I found informed me that Peekskill, New York, was founded in the late 1600s, when the Dutch trader Jan Peeck established a trading post at the mouth of Annsville Creek with the Hudson River. The trading post at "Peeck's Kill", was located at the far end of Annsville Creek on the shores of what is now Peekskill Hollow Brook. ("Kill" means "creek" in Dutch.) The Native American Kitchawank people called this location "Sachoes."

Because of the easy access to the Hudson, the city became a manufacturing center for the needs of the early settlers. First boats of all descriptions and later steamships stopped for passengers and cargo as they traveled the river from New York City to Albany. During the Revolutionary War, it was considered a strategic key to the defense of the nation, having twice been burned by the British. Colonialists John Peterson and Moses Sherwood of Peekskill instigated a rifle and cannon attack that helped foil the famous John Andre meeting with Benedict Arnold while a local citizen interrogated and captured the spy Andre at Tarrytown. Prior to and during the Civil War it was associated with the Underground Railroad. Henry Ward Beecher, the famous 19th-century abolitionist and cleric, lived in Peekskill. His daughter, Harriet Beecher Stowe, authored Uncle Tom's Cabin.

Mills and tanneries sprang up and because of a plentiful supply of local iron, Peekskill became a center of stove and plow production by the early 1800s. A blast furnace at Annsville Creek produced pig iron. The Annsville area also had a large wire mill and Captain Joseph Binney's Peekskill Chemical Works produced products that eventually became the foundation for the creators of Crayola Crayons. Farther down the Hudson, in the Croton River, was the Bailey wire and rolling mill that was the subject of my early historical review associated with Bailey McCarthy's and Nick's family history. Of particular interest in my look at Peekskill was that Frank L. Baum, author of The Wizard of Oz attended the Peekskill Military Academy. Rumor is when he asked upon his first arrival off the train how to get to the Military Academy he was told to "follow the road made from yellow brick" that led to the academy.

It was early summer, and I entered Annsville Creek by paddling from the Annsville Creek Preserve Park at Annsville Circle. It was still morning when I unloaded the kayak and got it over on the jet ski pier. The river fog was thick in parts as the warm river exchanged moisture with the cooler air above as my kayak slipped into the water. When I reached back to make sure I left the rope where I could grab it on my return, the motion almost flipped the boat, and I smiled that half-nervous-half-relief grin as I righted the boat. The way this particular launch was built, there were slots in the rubber with ropes attached and dangling just into the water. To exit the water, you had to grab the rope and pull yourself up on the rubber pier, all the time fighting to maintain your boat's balance until enough was out of the water and sable.

As I turned away, the view spread out across the small bay. The water was like glass, and I could see four mallard ducks crossing in front of the far bank as the Am-Trak commuter train flew across the low Metro-North Railroad Bridge over Annsville Creek, crossing this corner of the Hudson on its way upriver. As I paddled from the Hudson toward the mouth of the creek, I saw two snow-white swans lingering near the shoreline before the Annsville Circle bridge that connects Peekskill to points northwest toward Bear Mountain and West Point or toward Wappingers Falls and Poughkeepsie.

It was not long past low tide; the water was rushing into the creek as I maneuvered across the silted-in mouth with the bottom seeming to be less than a foot under my kayak. Many people don't realize that the Hudson is a tidal River Estuary of the Atlantic Ocean. Salt water from the ocean combines with freshwater from the river, and the streams and creeks that feed it with this "brackish" water extend from the mouth of the Hudson in New York Harbor 153 miles up-river to the City of Troy. As I paddled across the shallow depths, I noticed large fish rolling and thrashing in the mud to both my left and right. Only later when I saw the same thing in Tonawanda Creek near Buffalo did I learn that these were large carp mating.

At first, I thought of paddling north toward the bay to the oil tanks at Roa Hook and toward Camp Smith to the north, paddling toward the low underpass that goes to some salt marshes on Camp Smith property. But I knew this would be a quick trip, and I was much more interested in going up Annsville Creek. When I entered the little bay, the Route 9/6 overpass was directly in front as I waved to a little boy staring intently at me out the car window waiting at the light before it again motored on its way to Peekskill or Cortlandt.

As I traveled up the creek just past the bridge, I noticed that this natural waterway was scarred by a hundred or more years of industry. As is my want, I typically hug the shoreline, looking for interesting bits of history, old and new. I quickly became aware that the banks were lined with concrete and stone and industrial slag: the remnants of its industrial past. This was still visible on this early summer day whereas a few weeks from now it would be overshadowed by the now visible encroachment of large areas of phragmites and other invasive species that would no doubt obscure this manufacturing past with their thick mantel by mid-summer.

As I paddled up, I found remnants of the more natural riverbank that was reclaiming and covering up civilization. Looking in the clear water, I could see schools of smaller fish and a rather large snapping turtle. The creek forks, and I chose to take the south fork under North Division Street bridge into a much smaller creek. Once under the bridge, I passed by some industrial buildings. At this point, the creek was very clear, and I was looking for more carp and small fish.

Later that night, as I fell asleep thinking about my short trip up Annsville Creek and my future move to Buffalo, my eyes became heavy, and I began slipping back and forth from dream to consciousness until I was no longer sure which state I was in as the thoughts were designing a painting in my mind. Historical events began to intermingle with more recent ones, and scenes from Nick's ancestry became juxtaposed in my mind with his image at that point in his life. The deep dive I did in my previous work entered my mind and placed the image of his ancestor Abraham Bailey from the 1930s against his own from 1979. I woke up in a cold sweat suddenly just as the two mirror images began to overlap; my cloths and sheets were drenched.

Driving through the Catskills a few months later on a bright late fall day, winding along Route 17 with the streams passing back and forth, I started to reflect on Nick's life before he wound up in Buffalo. This was my first visit back home from Buffalo, where I had spent the first few months getting settled and finding my way. Gone were the scattered fly fishermen waste deep in the middle of the flow that I saw on my spring trip north along Route 17 just months before.

The now magnificent color of the fall played across my windshield view with every mountain range I passed. I was thinking how much I loved this ride no matter what time of the year. It was just magnificently beautiful, certainly as beautiful as the spectacular Hudson River Valley only in a much different way. The feel of the mountains and the fresh, clean air soothed my soul as I looked up at clouds seemingly fingertips away. The air breathed easy in the mountains. All of this cleared my mind for the analysis needed, which required me to reflect on life and how very great it is.

I pulled off at what was rapidly becoming one of my favorite stops in the Catskills: Roscoe, New York. Trout Town USA, the ultimate fishing town, was still more rustic than commercial, and it was home to five of America's top trout streams and of course the world-famous Roscoe Diner. Roscoe is located in the Southwestern section of the Catskill Mountains in an area of rugged mountains and pristine rivers, lakes, and ponds. Just west of town, the world-renowned Beaverkill River and Willowemoc Creek meet. This is the place where for hundreds of years fly fishing luminaries gave dry fly fishing the prestige it holds today. Farther west, the East Branch, West Branch, and the Main Stem of the Delaware River flows, providing anglers more phenomenal fishing.

During the drive, I thought about this dream and what I had learned so far. Nick McCarthy graduated from the University of Massachusetts cum laude in 1977 and proceeded directly to Yale University School of Public Health, a department of the medical school. He really enjoyed UMass, having gone there with his friend from high school Mike Twardy to wrestle. Mike and he had gone to wrestling camps together in high school at Le Moyne College after their sophomore year and Lehigh University after their junior year. Then both went to wrestle at the University of Massachusetts. They both were captains of the high school football and wrestling teams.

Mike was a superior wrestler, having gone to the states in high school and wrestling all four years at UMass. Nick had lost his match in the sectionals and never made the states. They lived in the same dorms until their senior year when they moved off campus. Nick wrestled through his sophomore year at UMass and then was medically exempt because it turned out he was born with one kidney, which he only found out during his athletic physical because his blood pressure was elevated. Truth is he was really done with wrestling anyway and wanted to concentrate on school and his girlfriend. He was tired of the coach's demands, including wanting him to spend his whole Christmas vacation traveling south to wrestle schools. Nick really just wanted to be home with his friends. Although Nick was a very good Division 1 wrestler, he saw no future in it and knew he had lost his desire to put the time in to wrestle internationally.

He did not want to go to Yale because he had other aspirations. But his mom said in no uncertain terms, "How can you turn down going to Yale after being accepted?" and so he went. He had applied to a series of graduate schools as close to his hometown as possible because he wanted to be close to his girlfriend. There was no thought in his mind that applying to Yale may have been a reach. All he knew was that it was one of the few schools nearby that had a graduate program in public health. But after his breakup with his girlfriend halfway through his senior year at UMass, Nick's mission was to put Peekskill and everything about it behind him.

His plan was to go to Vale, Colorado, and work at a water treatment plant that his friend who lived in Vale had set up for him. He had recently decided to bypass the assignment he got in the Peace Corps, the Ivory Coast, because he was not ready to give up two years of his life—although the idea did appeal to him. No, he had concluded that he had to work his life out a different way, and Vale was it. Then Yale came along and spoiled it. He had interviewed for graduate school right after his UMass graduation, not long after he broke up with his girlfriend.

He went to the interview still raw from the breakup with an attitude that could only be described as arrogantly nonchalant, which resulted in an air of confidence. This must have made quite an impression on these Yale interviewers, because the unlikely candidate received an acceptance letter to this prestigious university. His already strong background in public health from his undergraduate degree, which was a rarity in 1977, coupled with his total lack of uneasiness at the process—almost like he belonged—must have been the winning combination.

Although there were fifty or more students in the graduate program that year, there were only five others in his environmental health discipline. Most of the public health students were in the hospital administration discipline because it made more sense to the medical doctors in the program and those who were looking to actually make money. The other disciplines included epidemiology, biometry (health statistics), and infectious disease studies. This was still in an era where the pursuit of science was mostly untainted. It was many decades before former Portland State University professor Peter Boghossian claimed in his 2021 resignation letter, "the university has made this kind of intellectual exploration impossible. It has transformed a bastion of free inquiry into a Social Justice factory whose only inputs were race, gender, and victimhood and whose only outputs were grievance and division."

Science is NEVER settled and anyone saying so is not a scientist or at least an ethical one and anyone claiming to be "science" is not or no longer a legitimate scientist! Were they corrupted by big government and big business? In a journal article in 2022, two leading scientists suggested that a paper in a leading scientific publication was based on intentionally falsified research. They suggested that their findings raise the question of whether foundational papers for major contemporary regulatory policy that lack scientific legitimacy should be retracted and that their findings also should serve as the basis for considerable ethical concern, as well as a prompt for ongoing ethical diligence and rigor in the conduct and publication of scientific research. Their detailed reviews revealed numerous occasions of bias, error, intentional research misconduct, and self-interest by leaders of the scientific community of the 1940s–1960s era. This misconduct had considerable influence upon individuals, such as Rachel Carson.

Rachel Louise Carson (May 27, 1907–April 14, 1964) was an American marine biologist, writer, and conservationist whose influential book Silent Spring (1962) and other writings are credited with advancing the global environmental movement. There are unscrupulous people in every profession, but when those in the scientific community—these supposed ethical professional scientists—write highly influential papers that are reflective of biased publishing, which become pivotal to advancing an agenda, all is lost. One example of despicable ethical irresponsibility occurred during the pandemic of 2020 and the whole "green" movement of the early 2000s.

Nick was not the normal Yale student. He was neither from a wealthy family nor from an Ivy league background. Neither did he seem excessively smart. During his two-year stint in graduate school, he worked nights cleaning offices and during summer, Christmas and Easter breaks on the back of a garbage truck. This work, along with blue collar language skills and his strong jock-like appearance, made him instantly recognizable as different than his fellow students and heralded him as a different person than the snobs at Yale. Those upper-class elitists looked down on the lower classes.

When Nick would go home for holiday to work on the garbage truck, he often reflected on his experiences at Yale. As he stared blankly at the "hopper juice" and maggots swashing back and forth as he hung on to the back of the truck, his classmates were off sailing yachts around the Mediterranean Sea and through the Strait of Otranto, visiting the many islands in the Adriatic. The hopper juice was a murky gray liquid brew, frequently with a dash of orange or yellow streaming into it from the top of the hopper as it comingled to create an ever changing "sauce." Hopper juice was the resultant mixture from the hydraulic compression of refuse as it released its fluids made up of chemicals, liquefied food waste, the leftover contents of mayonnaise and mustard jars, and whatever else remained form containers of drugs, perfumes, toothpaste, soda cans, and the myriad other things humans toss in the trash.

To say he was somewhat out of place in this elitist academic world would be understating how the unsophisticated jock made the requisite impression on his classmates. But this impression and their somewhat reluctant acceptance of him did not prevent them from seeking his answers on take home tests once his academic prowess and past public health knowledge became clear. Nick thought it richly ironic when his friends from home would introduce him as "This is Nick; he goes to Yale," which was followed quickly by their quizzical look when they realized what they just said. Knowing Nick as one of them, they inevitably followed with the and question "How did you get into Yale?" Because when Nick was with them, he was one of them, an ex-dumb-jock and a fun-loving guy, certainly not an elitist snob or geeky member of the honor society.

The most important thing Nick learned from his parents, and perhaps it was passed down through many generations, was to live your life with a purpose and with personal responsibility. Despite these graduate school years being two of the worst and least fun years of his life so far, after a few days in New Haven he knew he was smart enough to be where he was.

On the very first day when he showed up in the public health building, he was ushered along with the rest of his class into a large auditorium. Naturally, Nick grabbed an aisle seat in the last row so that he had a perfect view of the whole theater that was about to commence, and an easy escape route should he become bored with the ritual. As the event progressed, he began to increasingly panic; his legs were ready to bolt out the door. All the graduate school public health advisors and a few doctors from the medical school were up on stage. The first speaker was the dean of the Public Health School who was going on and on about how they, the students in the audience, represented the cream of the crop, the future leaders of the country, future heads of public health departments across the country, future business leaders and pioneers. Wow, Nick began to think, "I'm in a way different environment here." He had never heard such talk from his normal friends or in high school and certainly not at "Zoo Mass" (University of Massachusetts) and certainly not from the guys at the garbage company where he worked every summer.

After these opening remarks, each advisor walked to the microphone to introduce his specialty and the objectives of their department. During each successive speaker, students in the audience would ask questions. This was not the average language of the University of Massachusetts student or kids from his high school class or his former teachers either; it certainly was not the language of the garbage truck. Nick could not understand half of the words that were being used, and he instantly felt that he was in way over his head.

Medical terms, hospital lingo, and statistical terminology were one thing, but the words being used by his peers were a vocabulary he had no experience with. Luckily, before Nick completely panicked, his own advisor started speaking. Because of Nick's undergraduate background in the field of public health, his advisor's terminology was very familiar. But what made the difference for Nick was when students rose to ask him questions. Nick realized that these students were using "big words" to ask simple questions. That was all Nick needed; he instantly knew he would be fine.

He knew the game these classmates were playing by using big words to say nothing. It was an experience he would have many times in his later career, and this lesson served him well. Before this, Nick's isolated world was populated with honest, hardworking, and simple people who, regardless of their intelligence, spoke in pure, simple, and humble styles—some of which could be rather salty and aggressive but certainly not refined or elitist.

As time progressed in graduate school, Nick became both bored and lonely. He was stuck in a one-room apartment in one of the worst parts of New Haven, Connecticut, in 1978 and 1979. He powered through the two years taking in as much as he could from the experience. He would spend hours at the gym, which he found to be a wondrous place to escape to and to maintain his athletic body. It was at least a city block in size and its ancient parts contained photos of all the past presidents and other illuminati that played at Yale, a wooden horse to practice polo, rowing pools, two swimming pools, rackets, and handball courts, and so on. He would spend hours working out, even playing pickup basketball, and was always hitting the speed bag alternating with jumping rope until he could not stand.

Nick loved people, but he was not too taken with his Ivy League classmates. Gone were his friends at UMass even with their quirky New England style, which was a strange cross between social liberalism and puritan conservatism. They seemed to be wary of people they did not know but enormously gregarious once they did. Most of the students at UMass came from the working class of Massachusetts because it was a state school. The closest Nick ever came to what he later found at Yale during his UMass days was when they wound up at some Smith or Amherst College party as reluctant Zoo Mass guests. Nick attended some of his more memorable parties at Smith College; they were full of opulence, fantasy, and elitism. And here he was at Yale with those very same people. Some were from Smith, and Nick had the inclination to say, "How you like me now, girls?"—but he didn't. His life so far had culminated in his adverse experience at Yale.

Nick found the public health course work very interesting. Although his major was environmental health, he was required to take courses in all the other majors, including hospital administration. He had already known that being a doctor was not something he wanted to be because he hated sickness and hospitals and death. This of course is another reason that his major at Yale was not hospital administration and all the course work and time spent in the medical school and among his classmates who were doctors going for their master's in public health (MPH) degrees just enhanced his disinterest in this aspect of his chosen profession.

He spent many a Friday night at the medical school happy hour mingling with the medical students, enough time to know they were a different breed as he longed to be home at Mutt & Jeff's with his high school and neighborhood friends. His hatred of sickness, hospitals, and death came to the fore many years later at his wife's death bed. The outcome of her more than two-year battle was clear. He whispered to her—not knowing if she heard—"I don't like watching you die because I know I cannot bring you back." They had said their goodbyes and made their peace when she was still alert. In the last hours, he reflected on how precious life was, how cruel life was for allowing his wife to die at such a relatively young age, that his kids would be without their mom, and what role he would have to take from here.

Chapter 4

THE FIRST HAZARDOUS WASTE TRAINING

"Jedidiah," said the women I had hired to do some historical research as she walked into the bar-restaurant for our meeting and reached out her hand in greeting. She followed up the greeting quickly by saying even before she sat down, "I found something that I think you will find very interesting as I was going through some old records kept by Ecology & Environment (E&E), the first firm Nick worked for." "Really!" I said with enthusiasm. I had not expected such quick results, and I could tell she had something she deemed pretty important. "Tell me: What did you find?"

I had hired a private investigation firm I used that specializes in searching through historical records and information about people who lived many years before. I hired this firm because they were based in Buffalo, and I had gotten an excellent recommendation from the firm I typically used that was based in New York City. Luckily, in Nick's time, recordkeeping was not uncommon. But I was about to get very lucky in my quest to learn about Nick's life and a glimpse into his early career.

The investigator's name was Sarah. She was probably in her early 40s if my estimate of women's ages was any good. I figured she was at least near that age, because she was one of the partners of the firm and had a good deal of experience based on her resume.

"I found sort of a picture diary with notes that Nick had kept during the early days of his career." I became more excited as I heard her describe what she found as we settled down waiting for our pre-dinner drinks to arrive. I had already ordered a Michters rye with one ice cube while I was awaiting her arrival.

Apparently, Nick had kept a detailed record of the first few years of his hazardous waste career, including a detailed chronology of all the sites he visited and some of his other work experiences. As she shared the documents and photographs with me, I could not believe my luck. This would make my job so much easier and give me an unplanned head start into the research that lay before me.

I knew that I could fill in lots of details about these places by looking at the old state and federal records. That would give me the intense technical detail, but this document would provide me with Nick's real-time experiences shared in his words and photographs. The next day, I started in earnest re-creating the steps Nick took in his early career. "My god," I thought as I flipped through the journal. "What a find this is! I must remember to add a big bonus when I pay the final invoice for Sarah's firm.

Nick walked into the lobby of the hotel in Naperville, Illinois, just about 30 miles west of Chicago and headed for the lounge area, where he figured he'd find some of his fellow trainers. Sitting down at a tabletop Galaxian game, which had become the favorite of the traveling group of trainers, was James B. Moore "Himself" deep into a game trying to break the previous record set by another trainer. Jim nodded to Nick's presence and then quickly returned to the game. This was your classic 1980s retro-shooter video game. The arcade game was developed and published by Namco and released in 1979. These games were rapidly overtaking the pinball games of Nick's youth and became an obsession with some of the trainers to fill in the time between training classes and nightly drinking.

Nick had received his 40-hour hazardous waste training certificate in October 1980, just two years after graduate school. His certification read, "This certifies that Nick McCarthy has completed the 5-Day Hazardous Waste Site Investigation Training Course. Presented by the Field Investigation Team (FIT) National Project Management Office of the Field Investigations of Uncontrolled Hazardous Waste Sites Project." The 1980 course was one of the first ever E&E HAZWOPER training courses.

The company was given incentives by the EPA, bonuses so to speak, that they would receive if they met or exceeded certain milestones. Their first milestone was to set up, equip, and staff regional offices collocated with the 10 EPA regional offices. The next and almost simultaneous milestone was to develop a series of field operating procedures to be adopted as official EPA procedures designed for the proper investigation of hazardous waste sites. These included procedures for site safety and operational planning, chemistry, toxicology, hazard assessment, site characterization, personnel protective equipment, respiratory protection, medical monitoring, site assessment techniques, decontamination procedures, field surveillance (AirToxics) and sampling equipment operation—including the proper use and maintenance of real-time chemical and radiological monitoring/survey equipment—and others. These were to be used to meet the final milestone, which was to conduct official hazardous waste site training classes that were designed to be 40 hours and prepare the new staff and future workers for hazardous waste site work.

Only weeks before leaving office, President Carter in 1980 signed the Comprehensive Environmental Response, Compensation, and Liability Act, which would become commonly known as the "Superfund." The $1.6-billion appropriation, funded in part by taxes on petroleum and other chemicals, was intended for the cleanup of toxic-waste sites and oil spills.

Just prior to Nick being hired at E&E, the company had received the EPA contract in 1979 to establish Technical Assistance Teams (TATs) to support the agency's Oil and Hazardous Substances Spill Emergency Response program. The company was required to open offices in Washington D.C., as well as the nine other cities where EPA had regional offices and was just establishing the first eight training courses when Nick was hired.

Once the second contract was awarded, Nick worked on the huge tasks of getting the Field Investigation Teams (FITs) set up and prepared to investigate uncontrolled and abandoned hazardous waste sites across America and its territories. The mission was different than the TAT contact objective to develop rapid response teams or emergency response teams (ERT) for chemical and oil spill response in that the FIT objective was for investigation of Uncontrolled Hazardous Waste Sites. In 1986, these procedures developed for the TAT and FIT contracts formed the basis of the formal OSHA HAZWOPER operations and training procedures formulated in OSHA 1910.120.

Despite Nick's young age and being less than a year out of Yale Public Health graduate school, he found himself right in the middle of working on all these procedures along with the more experienced staff from across the country that had been recently hired. His training in public health coupled with the rapid experience and knowledge he had already received at E&E made him a perfect member of the training staff. Because of his role in the Buffalo headquarters office, Nick had been involved in a more demanding role.

He was intimately knowledgeable of many of the aspects of each milestone met by E&E. He had sat in the corporate conference room for weeks in Buffalo with Frank, a colleague who was part of the human resources and marketing department. Their assignment was to go through the thousands of resumes that were pouring in from around the country and categorize them by region, background, and company/project fit.

There were engineers, geologists, environmental scientists, industrial hygienists, toxicologists, chemists, and others, including many who were ex-Army chemical/EOD Corps and Navy nuclear scientists. Plus, there were experienced members of the hazardous and oil spill response teams from the Coast Guard. There were experts in from academia (internal/occupational medicine and toxicologists) supporting all this.

Nick was also given the task of finding and ordering all the equipment to outfit the twelve regional vans and regional offices to properly investigate hazardous waste sites across the country. The company had ordered 12 bread trucks to be retrofitted as response and site vehicles. The one thing that always perplexed Nick was the fact that none of these were 4-wheel drive off-road, which Nick always viewed as a major flaw. Obviously, they were ordered by someone who did not do much field work. "Oh well," Nick thought at the time, "at least they will get them close to most of the sites."

One of Nick's fondest memories was the day a large wooden crate arrived at the E&E Sugg Road Cheektowaga, New York, office and was off-loaded in the garage. Upon its arrival, Nick and two more senior staff, James B. Moore "Himself" and Boyd, walked into the garage and assembled around the large crate almost like kids on Christmas Day waiting for permission to open their presents. This present was full of non-sparking tools that Nick had ordered to outfit each regional office to go into the twelve panel vans that had been bought and outfitted as remote workstations and equipment storage units.

Boyd picked up a crowbar and pried open the lid, the nails screeching as they were yanked rapidly from their deep seats in the wood. Boyd was a big man and had little trouble yanking open the lid, which made the sound of the nails leaving the wood all the more accentuated. He threw his large right hand into the sawdust-filled box, his eyes getting big as he pulled out a bronze and copper-beryllium tool.

He exclaimed as he held the tool high above his head, "Behold the golden tool!" Each man quickly reached into the wooden chest, and soon twelve of each tool were lying in rows on the garage floor. There were 55-gallon drum bung openers and drum plug wrenches, hammers, adjustable wrenches, socket kits, and more.

The purpose was to supply each regional team with a set of tools to open drums and other containers of unknown waste. They had to be made of non-sparking materials to provide protection against fires and explosions in hazardous waste environments where sparks might ignite flammable solvents, vapors, liquids, dusts, or residues. They had done much research before Nick ordered these tools, because most of the work would be at uncontrolled hazardous waste sites where every container was an unknown, even if it were labeled.

By the time of this Chicago training course, Nick was a junior member of the National Training Program (NTP) team and had already received hours of training on the equipment and procedures he helped develop. He had also been out on a few hazardous waste sites across the country. This training course in Naperville was held in the Summer of 1981 and was just one of a series of other courses given in central locations across the country for the regional staff. The last of the initial courses was given in Denver, Colorado, in June 1982.

The courses were given jointly by E&E and the firm CH2M Hill, which was a partner on this contract with the EPA. It is common on larger government contracts for two or more firms to form a bidding team to strengthen their opportunity of being selected; each brings some advantage and fills in gaps the other may be lacking. Many times, on the nationwide contracts, teams were formed because one team member had a larger presence in one region of the country and government agencies liked to "spread the wealth," giving teams with multiple companies an advantage.

The training curriculum involved a 5-day, 40-hour emersion into the field operations basics needed to officially work on USEPA hazardous waste projects, and these courses became codified in the OSHA'S 1910.120 (HAZWOPER) standard requirements. OSHA's goal with 1910.120 was for it to serve as procedures so that no one would be injured or have their health affected as a result of working with or near hazardous waste. The procedures developed by

E&E/CH2MHill, NIOSH/OSHA, USCG, and USEPA Working Group established the primary OSHA requirements in the training.

They included the following specific topics:

• A robust Corporate Safety and Health Program defining the administration, key staff, and elements required to perform hazardous waste work, including a medical monitoring program based on toxicology of industrial chemicals;

• Site Characterization and Analysis training designed to educate staff on the potential site hazards, the characteristics of the hazardous waste, and the safety and health control procedures needed to protect employees from those hazards at each site;

• Site Control Procedures, which defined what employees needed to have for each site. These minimally included creating a site-specific site map; identifying site-specific work zones; establishing a buddy system; establishing the means of site communications, including an alert system for emergencies; establishing site-specific variations to standard operating procedures and safe work practices; and identifying the nearest medical assistance;

• Required Training defining that employees who may potentially be exposed to hazardous substances, health hazards, or safety hazards must be trained before they start work;

• A medical surveillance program, including tests to ensure employees can safely wear respiratory protection and other personnel protective equipment while at the same time creating a baseline blood/urine profile should future exposure occur as well as annual or post-exposure medical testing requirements;

• Engineering controls, work practices, personnel protective equipment (PPE), or a combination of these for protecting employees from hazards;

• Developing site specific safety and health plans designed using standard operating procedures for site specific hazards;

• Training on the proper use and maintenance of industry specific real-time monitoring devices to verify that control measures are effective and that workers are not being exposed to hazards at a level that exceeds the permissible exposure limits (PELs) or upper and lower explosivity/flammable conditions as work is progressing in real time;

• Procedures for handling and sampling drums and containers;

• Decontamination procedures for all aspects, including accidental exposures; and

• Written emergency response contingency plans for anticipated potential emergencies.

In March 1982, Nick attended special training on field operations and specifically the maintenance and operation of the HNu Systems PI 101 Photoionization Detector and the Century Systems (Foxboro) Model OVA-128 Organic Vapor Analyzer. These were the two main real-time organic vapor analyzers that existed in 1980, and Nick had purchased one of each for each regional office and the corporate office in Buffalo. The bulk of the training course material was prepared by two brothers who ran the operations in the Boston office, both E&E Field Investigation Team Leaders in Region 1 (New England). One of the brothers, Paul, also gave the subsequent training.

Much of the technical information was provided by Foxboro-Wilkes, the company that manufactured the OVA, and by HNu Systems Inc., the company that manufactured the HNu. These two instruments were the main real-time monitors used by the industry for many succeeding years before other instruments became more common.

This intense training lasted a week, and only two people from each regional office were selected to attend. Nick was one of several to attend from the corporate office because he was one of the newly formed hazardous waste team members that functioned out of the corporate office and acted like a stop gap, providing assistance for any regional office that needed it.

This team was also involved in all the hazardous waste and chemical and oil spill responses for the Western New York projects. Because of Nick's background, he was also part of the corporate health and safety staff and was sent out to the regional offices to complete audits of operations. In this role, he was also responsible for reviewing health and safety plans and maintaining the medical monitoring records companywide, which required Nick to sign a confidentiality statement to keep employee records secure. Nick spent many hours reviewing medical exams to ensure that staff was not performing duties they shouldn't based on medical issues such as hearing loss, elevated blood counts, or other physical limitations.

The OVA and HNu air characterization instruments were integral to the site characterization and health and safety monitoring performed during onsite work.

The weeklong training was designed to instruct the select staff in the operation of these complex instruments and required specialized training. At that time, only individuals trained in the use, operation, and maintenance of these instruments were authorized to use them at sites. Nick noticed later in his career, as years went by and other companies became engaged in this work, that this equipment-specific training and experience requirement went by the wayside.

Often, years later, Nick would find many people using the instrument with little to no knowledge of its workings other than how to turn it on and read numbers. They had no knowledge of the relative responsiveness of these instruments to various chemical compounds and organic vapor, and therefore they did not really understand what the numbers on the dial meant, creating hazards for themselves and others.

Each of the two instruments operated on separate and different principles. The OVA was a gas chromatograph with a flame ionization detector (FID), and the HNu was used to detect various organic vapors and used a photoionization detector (PID). The OVA employed the principle of hydrogen flame ionization using a diffusion flame of pure hydrogen and air. When a sample of organic material was introduced into the flame, ions formed, causing the flame to become conductive. The OVA could be used in the survey mode, which made the instrument a total hydrocarbon analyzer to monitor the total amount of ambient organic gases and vapors.

All readings from this instrument were noted as methane-equivalents because the instrument was calibrated to methane. The HNu was calibrated using the 10.2 ev lamp to benzene. It used other energy lamps to be sensitive to a broader range of organic chemical compounds. Course instruction and practice sessions provided a list of compounds that each instrument could detect along with the relative response for each compound of interest. The concept of relative response was key to the function and use of survey instruments because no field survey instrument at the time detected a specific compound at its real concentration.

In other words, these instruments did not "see" each compound on a one-to-one basis. For example, the OVA calibrated to methane would "see" 100% of methane in the air but would only see 64% of propane, 90% ethane, and 10% of carbon tetrachloride. Someone not understanding this concept of relative response could under- or overestimate the actual levels of a hazardous substance in the air. The same principle of relative response existed for the HNu and any other PID used today. However, because of their different operating principles, PIDs (such as the HNu) and FIDs (such as the OVA) detected different compounds at different relative responses.

In the gas chromatograph (GC) mode, the OVA functioned as a completely portable, self-contained gas chromatograph for gas or vapor samples. In the survey mode, both the OVA and HNu only detected total organic vapors because there was no way to distinguish which organic chemical or mixture of chemicals was causing the response. Therefore, these instruments were really survey instruments that were used by health and safety to require exit from a workplace or a change in the required level of protection based on a pre-established action level—the number at which a specific action, such as exiting the site, was to be required.

The problem with using the OVA in the GC mode at a hazardous waste site was that those were not ideal conditions for a GC such as would be found in a laboratory, because it was almost impossible to keep the GC at a constant temperature. Thus, results were not consistent and would change as the column's ability to function changed with ambient temperature changes. Over the years, Nick became very accomplished in the use of both instruments. He often used both simultaneously to assess different factors and gain more field data. He understood their deficiencies well as well as what the data really meant. He even had the opportunity to use the OVA in the GC mode on one site on which they set up a portable lab in the site trailer and used a constant heating element to keep the GC column constant.

The positive-pressure, Self-Contained Breathing Apparatus (SCBA) alarm was ringing as the trainee was just approaching the decontamination line, and Nick could see the panic developing in the person's eyes. The trainee knew what the alarm meant—that he was running out of air—and so did his teammates and the decontamination crew.

It was bad enough that many have to overcome the overwhelming sense of claustrophobia when dressed in a full protective suit with two layers of gloves and boots and having a full facemask covering the face as well as a hood pulled over to its top and taped in place. Heck, every joint was taped. Nick could see the increased anxiety overcoming the person as the fear of running out of air before completing decontamination and getting free of the suit was rapidly sinking in.

Nick had set up the command post and laid out the decontamination station in the back parking lot of the hotel just down from the conference rooms that they had set up for the training. This was one of Nick's main functions as the junior member of the training team because he was not quite qualified or experienced enough yet to be giving the topic lectures.

Each 40-hour course had a field exercise portion that required trainees to get fully dressed in various levels of protection, go "downrange" to fulfill certain field functions, and then return through the decontamination line in proper sequence. The exercise was designed to get new field personnel used to functioning as a team in various levels of protection.

It was also designed to reinforce the proper dress-out (donning) and dress-down (doffing) procedures as the group exited through the decon line either to go downrange or when coming back from the field. It was all based on the procedures taught earlier in the 40-hour class and was the time to impart the proper sequence of putting on gloves and clothing and taping and untapping joints to prevent exposure—both before site entry downrange and while getting undressed from the protective gear.

The training staff had purposefully designed the exercise with one tank short on air so as to gauge the team's performance under pressure and as a learning tool on what to do in such a circumstance. Once safely out of gear with the panic retreating, the main lesson to the team was how to gauge risk and consequence. Yes, preventing exposure to toxic chemicals was the main purpose, but reacting to an immediate life-threatening situation always trumped all else.
In this case, losing one's life due to suffocation, or being overcome by heat or

cold, or reacting to a medical emergency certainly overrode following exact proper procedure. The point of practicing these emergency situations was to know the correct procedures to follow even in the panic/stress of an emergency.

Chapter 5

AUDITS AT MIDCO I AND MOTCO TOO

There was a photo on the wall in an office of three men in Level-B chemically protective suits with self-contained breathing apparatus (SCBA) tanks on their backs. They are standing in a sea of 55-gallon drums that appear this way and that way at all angles and some stacked two or three high. There is no way to tell who they are or anything about them because the face shields obscure their faces and the protective clothing masks their physical features. Oh, you can see a difference in size and if you are not a casual observer, their authority, but that is about all. One, dressed all in green, holds a glass thieving rod for obtaining liquid and viscous samples from the drums. There is a blue/green material oozing out of its bottom into a sample bottle.

• • •

The rapid pace of Nick's early career at E&E provided him with almost a meteoric advancement in knowledge and experience. It didn't hurt Nick's development that he was a sponge absorbing everything he could. Because of his upbringing, experience in highly competitive sports, and personality, he understood the important dynamics of team interactions with different personalities.

In one short year Nick had gone from working directly for the president of E&E on his pet projects to being a major player gearing up for the first ever EPA contract to investigate uncontrolled hazardous waste sites across the country. In those early days, he sat in rooms with the most experienced people in this young industry developing the procedures that would be used by others for years to come. He was learning at the knees of the individuals who were paving the way in the early days of hazardous waste spill response and investigation. One such individual was Mac.

About a year into the EPA-FIT contract, once offices were set up and equipment in place, Nick had been transferred into the newly reorganized health and safety department at the Buffalo E&E headquarters office. One of his early roles was to perform office and site safety surveys of several offices and hazardous waste sites to ensure that the procedures were being followed consistently by every regional office. Not being a stupid person—and having enough wisdom in his youth to understand the nuances of his age and position—he was very cognizant of how to keep his burgeoning ego in check when it needed to be. Traveling to offices to "audit" individuals who had much more experience and knowledge than he had could have been a disaster for anyone with less finesse or less self-awareness. Luckily, he had some experience in a similar role when in graduate school. For his thesis on foodborne illness, Nick took an internship as a food/restaurant inspector. His job was to go into various restaurants that were preparing foods for summer camp programs run by the State of Connecticut.

Nick learned early on that many of these inspectors were for the most part treated with suspicion by the staff and owners. What Nick picked up on quickly was that either many inspectors were "on-the-take" to overlook things or were complete assholes drunk on authority and power. Nick's approach was to put the people at ease and quickly show them that he was not looking for money and certainly was not going to be an authoritarian jerk. Authoritarian jerks—you certainly know the kind—give offense in some superior or authoritarian manner either without a shred of awareness that they're doing so or on purpose because they have the "power." The sad truth is, everyone can be a jerk sometimes, and during his restaurant inspection days Nick was conscious not to be. This lesson was well used by Nick in his new role of an auditor.

The Technical Assistant Team (TAT) leader in Chicago was Mac. Nick's trip to the hazardous waste site MIDCO I in Gary, Indiana, resulted in one of the more iconic photographs of early hazardous waste work. It was a photograph of Mac and his team surrounded by 55-gallon drums of unknown content, wearing Level-B personnel protective equipment (PPE) consisting of a chemical protective suit, disposable booties, chemical resistant gloves, and a positive-pressure SCBA. Mac knew that he had to work quickly since there was only 30 minutes (or less) of air in the air bottle on his back. Mac had carefully removed the bung plug from the top of one of the drums and screened the head space in the drum for flammable vapors prior to collecting a representative sample of the drum's contents. A single-use glass rod thief for obtaining the representative sample protruded from his hand. There was blue/green material oozing out of its bottom into a sample bottle. Since the SCBA only allowed for 30-minutes, many trips into the "hot zone" had to be made before the "sea of drums" was categorized. Later, multiple teams and variations on the supplied air set up was used to increase the efficiency from this initial assessment phase.

Nick had met Mac at several of the training venues and meetings because he was one of the more experienced chemical spill response/hazardous waste workers. Mac had received his training and gained experience as a member of the Coast Guard and when he worked at the New Jersey Department of Environmental Protection, and Mac instantly impressed Nick when he first met him as someone who obviously "knew his stuff." During the trainings, Nick was always interested in the various real-life experiences Mac would use in his training classes and some of the unique tricks he would impart. One such trick was using a blow-up beach ball to plug a deep sewer intake.

Typically spills occur at a facility, and it's always necessary to reduce the spread of the spill as quickly as possible. Later in Nick's career, he would see a number of spills flooding into the nearest storm sewer, releasing their effluent into receiving ditches, streams, or rivers. As Nick learned during these early training courses, the priority and first step at chemical and oil spills was stopping the spill and its spread as quickly as possible. It's hard to explain the adrenaline flow you have when you arrive at a spill scene where all hell has broken lose and you need to keep your wits about you and get things in motion almost instantly.

The first step is to quickly assess the spill scene and the lay of the land. This is followed almost instantly by action. Typically, the first emphasis is on characterizing the hazards, setting up operating perimeters, and trying to stop the spill from entering into culverts, drains, and ditches. Concurrently or shortly thereafter, the spread of what has already escaped must be stopped as much as possible. Once those tasks have been accomplished, the long and tedious job of cleaning up the spill commences. Mac's lessons often made the point that the decisions made during the first few steps greatly affect the time required to complete a cleanup and its eventual cost both monetarily and to the environment and wildlife.

Nick's first thought as he was sitting in the plane from Buffalo to O'Hare Airport in Chicago was how in the world would he, Nick—still young and with limited experience—audit someone like Mac. From his past experience, he had an inkling of the tack he must take. That was to be low key, nonaggressive, and complimentary. He viewed this as a learning experience, and that was the exact impression he wanted to make from the outset. Sure, he would review the office records and watch the field procedures because that was truly a necessary objective of the safety survey, but he had no intention of leaving the impression that he was there to tell these guys anything about their business. It was a delicate dance but an important and necessary function.

The schedule was to meet Mac and the Field Investigation Team (FIT) and TAT team in the regional office on Day 1 to review the office layout and equipment. The next day was to observe a field job Mac was running at the MIDCO I site in Gary, Indiana. The office audit went off without a hitch. As would be expected, everything was military ready and in almost perfect order—procedures, records, equipment, instructions, etc., were all in their proper places in file cabinets, and the equipment was stored away clean and in proper order.

When Nick first arrived, he was taken to Mac's office. "So nice to see you again, Mac," Nick said as he reached out his hand in a friendly gesture. As they were shaking hands, Nick continued, "You may remember I was part of the training team from Buffalo, and I made it a point to sit in on your lectures the days you were there. I really learned a lot listening to you, and I am looking forward to watching you in action tomorrow. Sounds like an impressive field job, and I am hoping to come back to Buffalo with some great photos and of course more knowledge." Nick's approach for impressing Mac—showing that he was there to watch and learn and not really to audit—was the correct one, as Nick intuitively knew it would be.

In fact, it really was the truth. Nick went on as he sat down: "Part of the EPA contract requires us to audit the regional offices and provide records to the EPA Washington to meet certain milestones. The company contract fee is partially dependent on meeting those milestones." Mac's demeanor was noticeably more at ease after that opening exchange. He said, "Happy to have you guys from the corporate office come out and see what happens in the field." Mac definitely had the air of someone with confidence who had spent time in the military. The rest of the day was spent meeting various team members from both the TAT and FIT offices, and Nick had a chance to briefly stop by and say hi to both Joe Petrelli, the FIT Team Leader, and the Regional E&E Administrator, Edward Lee Howard. Howard would later gain great notoriety as described later in Nick's story.

Prior to this visit, Nick had a number of conversations with Howard because one of his roles in Buffalo was to manage the "special project fund" that FIT/TAT leaders would draw from when they needed things not in their budget. That night, his first ever in Chicago, he went out on the town with some of the FIT staff he met at training courses. They went for dinner at some famous place and got some deep-dish Chicago style pizza and then caught a Chicago White Sox game after a few beers at downtown bars. However, Nick was very much looking forward to seeing the famous MIDCO I site the next day and watching Mac in action.

MIDCO I, the very first hazardous waste site remediated under the recently enacted Comprehensive Environmental Response Compensation and Liability Act (CERCLA), or Superfund, was one of the more impressive hazardous waste sites Nick was involved with to date. By the time his career was over, he would be on many more. To prepare for the site visit, Nick read the EPA Superfund records to gain an overview of the site.

Nick was used to consuming the technical detail that was in the EPA documents. The overview of the document read: "The Midwest Solvent Recovery Co. (MIDCO) I site occupied four acres in an area of wetlands in Gary, Indiana. In April 1975, the company began storing and reclaiming thousands of drums of hazardous wastes on site. Reports indicated that the company also dumped sludges and other wastes into a pit on the property. In December 1976, a fire destroyed more than 14,000 drums, essentially halting operations. In late 1977, operations started up again and stopped in 1979. Several thousand drums containing materials such as paint sludge, solvents, acids, caustics, and cyanides were left on site, many of them deteriorated and leaking. The drums burned in the 1976 fire also remained on the property. Surface water, ground water, and soils were contaminated. In June 1981, the EPA fenced the site." This is how the site was when Nick showed up to view Mac and his team in action.

When Nick arrived at the "command post" set up at the site, Mac and his team were already downrange in full protective gear. Nick examined the decontamination setup so as to be sure to have notes for his report. Of course, it was precisely set up and ready to receive the downrange crew as they left the site; the decontamination tubs, soap and water, and safety kit were all arranged in textbook fashion. Nick found a good vantage point from which to observe the action. He had brought his recently purchased personal Minolta X-700 35mm camara with a Nikon 70-300mm f/4-5.6G AF Telephoto Zoom lens. Looking through the lens, he could see the actions of the field crew up close. Mac was dressed in green, and the two crew members were dressed in yellow. Nick took a series of photographs, one of which was the iconic photo of Mac taking samples. Many years later, Nick came across Mac while he was visiting the URS's Buffalo office. URS was a large, National engineering firm. Mac had left to work at their Buffalo office which was run by John C. Gerter who grew a small office to one of the largest in Buffalo, NY. When Nick heard Mac was working there, he made a point to visit Mac's office. On the wall was a blown-up copy of the photograph.

After Mac's work and subsequent investigations, an interim corrective measure was completed at MIDCO I. The surface wastes and drums, which also included an underground tank, were removed along with the top foot of contaminated soil. The primary contaminants of concern affecting the soil, sediment, and ground water were identified as various volatile organic compounds (VOCs), including benzene, toluene, and trichloroethylene (TCE-chemicals used to make refrigerants and as a metal degreaser.)

Other organic wastes included polychlorinated biphenols (PCBs), and polycyclic aromatic hydrocarbons (PAHs). PCBs were used extensively by electric utilities as noncombustible dielectric fluids in transformers and electrical cables, and by natural gas transmission companies as noncombustible upper cylinder lubricants and gas turbine lubricants. PCBs are a suspected carcinogen at low concentrations and are now regulated under the Toxic Substances Control Act (TSCA). Earlier in his career at a site survey at the Roebling Steel site in New Jersey, Mac determined that a fire retardant that coated a corrugated metal wall may contain high levels of PCBs. A sample of the fire retardant was found to contain over 20% PCBs.

PAHs are a class of chemicals that occur naturally in coal, crude oil, and gasoline. They are produced when these products as well as wood, garbage, and other fossil fuels are burned. PAHs generated from these sources can bind to soils along with metals such as mercury, chromium, arsenic, and lead and become health hazards when people come into contact with the soil or dust.

During the operations at MITCO I, wastes were dumped and spilled onto and into the ground at the site. The fire that destroyed thousands of drums resulted in additional spillage of chemicals onto the site. The contaminated ground water and the subsurface soil and debris below the top foot of soil was not remediated until long after Mac's and Nick's time on site. A Remedial Investigation/Feasibility Study (RI/FS) was later completed under the EPA and the Indiana Department of Environmental Management (IDEM) oversight from

1985 to 1989. This study showed that subsurface soils, both fill and the natural soils below, were still highly contaminated by a large number of hazardous substances. The fill contained some cinders and gravel mixed with lots of debris, including crushed drums, paint waste, wood, concrete, bricks, and other materials. Ground water below the site was found to be highly contaminated. After Nick's and Mac's time on the site, MIDCO I was listed as an active National Priority List (NPL) superfund site by the EPA and was considered one of the worst hazardous waste sites ever identified by the organization.

About the same time Nick was auditing MITCO I in Gary, Indiana, his supervisor Dave, E&E's Corporate Health & Safety Director, was auditing the MOTCO Inc. site in the City of La Marque, Galveston County, Texas. Galveston County is located on the Gulf Coast of Texas eighty miles southwest of the Louisiana state line. East of Brazoria County and west of Chambers County, it is bounded by the Gulf of Mexico to the southeast. Not far south of Houston, La Marque is known as the gateway to the gulf and the hub of the mainland.

Dave was an experienced person in chemical spill and hazardous waste work, having been in charge of the TAT-Emergency Response Team (ERT) office in Cincinnati, Ohio, before transferring to Buffalo to be the corporate Health and Safety (H&S) Director.

The hottest days of the year are typically late July and August, and temperatures in Galveston County typically range from 80°F to 90°F. "Could this be the muggiest day of the year?" Dave thought as he stepped out of his rental car, the humidity nearing 91% with little to no wind. Dave showed up on this mid-summer August morning to this south Texas site to complete a site safety audit of ongoing operations. Already onsite were a very large group of team members from the USEPA and E&E regional team attired in various personnel protective levels ranging from levels B and C to D.

For those not in the business, the levels of personnel protection range from highest and most protective Level A to B to C and lastly D with variations applicable to each depending on the job-specific chemical hazards and field conditions. These were all formally and extensively spelled out in the detailed TAT and FIT field procedures and all field workers were trained on these during their mandatory 40-hour training course before they were allowed onsite. Dave could tell by their demeanor that this was not this crew's first rodeo because they appeared to be functioning as a well-oiled and seasoned crew.

Level A is the highest level and typically involves a fully encapsulating heavy chemically resistant suit, which in these early days were made of Butyl Rubber to protect workers from dust, gases, and splashes. These level A encapsulation suits had vapor-tight seams to provide the highest level of skin protection and had an expanded back to accommodate a SCBA tank and harness.

The suits also included chemically resistant socks and gloves or glove liners, integrated face shields, and hoods to make them fully encapsulating. Many times, a second or third pair of gloves and boots of different rubber types were worn over the suit to increase the chemical resistance depending on how the protective rubber reacts to certain chemicals. In unknown conditions with unknown chemicals, it's important to protect against a wide range of chemicals with varied chemical and physical characteristics. Between the multiple layers of face shields, gloves, and boots, Dave and Nick often wondered whether the hazard from physical trips, falls, and reduction of sight angles and manual dexterity—as well as the potential heat stress—was a greater, certainly acute hazard than the chemical exposure.

As Dave walked along the access road leading to the site, the scene opened up before him and enveloped his mind as he quickly scanned the multilayer scene: almost a moonscape of brush and pits and lagoons and groups of people in various locations hard in the action of site hazardous waste work. He intuitively gazed past the trailer that was established as the command post as well as the decontamination zone beyond and focused at the first thing that caught his eyes.

He dismissed the nondescript because his eyes focused quickly on a group of figures dressed in Level B with their SCBA's slung ackwardly over their backs. They were busy sampling soils and materials near what appeared to be black open-pit lagoons filled with all kinds of tarry, resinous industrial wastes. His eyes then shifted to another set of workers farther away from the first crew dressed in Level C personnel protective clothing, which included chemically resistant coated Tyvek suits, chemically resistant gloves and booties, and full-face air purifying respirators with cartridges to filter the outside air they were breathing. This crew was busy sampling groundwater from some perimeter groundwater monitoring wells, while using a PID to monitor for organic chemicals in the air. As he swung his head around, back toward the trailer he noticed some other crew members now more recognizable dressed in Level C stationed at the decontamination stations. Even more fully recognizable crew members were dressed in Level D attire, which consisted of steel toed and shanked safety work boots and work cloths; some were wearing surgical gloves, but there were no face-disguising shields to blur their faces. This team was busy in non-contaminant areas, where personnel protective clothing was not required.

Dave proceeded to the command post trailer, and as he opened the door he was greeted by the E&E regional office safety officer who was also functioning as the designated site safety officer on this project. He was on his way toward the decon line, telling Dave as he scurried behind that he was monitoring the onsite personnel for signs of heat stress. Also the SCBA user time in the field was getting close. Just then as a distress signal came from the downrange crew. SCBAs usually contained enough breathing air for about 30-minutes or less depending on the stress and the worker's experience in their use. The EPA's Onsite Coordinator was also tagging along because his job was the overall direction of field operations.

Dave noticed the worried expressions on their faces; there was obviously something wrong. Suddenly they all came to a halt on the "clean" side of the decontamination line. The EPA coordinator was saying something about what he was just told over the walkie talkies by the downrange site leader.

Besides the EPA and E&E teams, Dave just noticed another set of contractors in Level C personnel protection on the far side of the styrene tar lagoon positioned in a small jon boat. The jon boat was a small flat-bottomed boat constructed of aluminum with one bench seat. They are typically used for fishing in calm waters. Attached to the bow of the jon boat was a rope that reached completely across the lagoon to where other members of the contractor crew stood by an old pickup that had its left-rear tire removed and replaced with a wheel rim, which they were using sort of as a pully or more like a "cat-head" that exists as the most dangerous part of a drill rig.

The idea was to obtain samples of the tarry contaminants and to try to judge how deep the lagoon was in places. The lagoons were crusted over with styrene tar, below which were solvent, dense nonaqueous phase liquids (DNAPLs) and nonaqueous-phase liquid (NAPL) compounds. DNAPLS and NAPLS are organic liquid contaminants that do not dissolve in, or easily mix with, water, such as oil, gasoline, other petroleum products and solvents. They tend to contaminate soil and groundwaters for very long periods of time and are known as persistent organic pollutants. Many common groundwater contaminants, such as chlorinated solvents and many petroleum products, enter the subsurface either through spills, leaks, or active dumping in nonaqueous-phase solutions. They do not mix readily with water, either on top, or deeper in the groundwater aquifer.

Dave noticed that the "crust" on these lagoons was not hard enough to walk on safely, so these contractors were using the jon boat to go out and test/break the surface. But they were doing so in the totally wrong protective clothing in an unsafe manner using unsafe procedures. The first rule of respiratory protection, for example, is that you must use a type sufficient to protect against the hazard.

Decades later this rule was not followed by the supposed "experts" in charge of the covid pandemic in 2019–2022. Under OSHA, you are prohibited to wear a Level C air purifying respirator if you don't know the nature of the chemicals, their concentration levels, or warning symptoms. None of these characteristics could possibly be known by these contractors in the middle of this chemical soup while they were churning it up and causing it to emit higher concentrations into the air. By contrast, the EPA/E&E crew downrange in the same area were wearing Level B self-contained breathing and personnel protective protection. This is what suddenly caught the attention of the EPA/E&E site crew who communicated it back to the command post as an extremely unsafe working condition.

The MOTCO Inc. site began in 1959 to 1961 as a styrene tar recycling facility and a state permitted disposal site for petrochemical wastes from nearby refineries. An industrial disposal facility operated on the site from 1964 to 1968. Today it is a 12-acre field that lies along the east side of I-45 right before you head over the causeway to Galveston Island.

For several years, millions of gallons of toxic waste flowed into seven unlined pits/lagoons. In 1968, the stench from the toxic brew was so bad that the City of La Marque shut down the operation. In the 1970s, the site changed ownership. One participant was a Minnesota Company called MOTCO, which planned to recycle the waste. But the project was eventually abandoned and went bankrupt. In 1982, the MOTCO site was ranked the most hazardous Superfund site in Texas.

A later 1985 EPA.gov site assessment report estimated "treatment of up to 15 million gallons of water is expected to be required during the cleanup. Under the water, there exists seven million gallons of organic liquids of which five million gallons will require disposal in accordance with the Toxic Substances Control

Act (TSCA) due to the presence of (PCBs). Under the organic liquids, approximately 18,000 cubic yards of sludges and tars are present. Under the sludges and tars are an estimated 45,000 cubic yards of highly contaminated soil." Much of this information was generated by the initial work of the EPA/E&E team and subsequent investigations.

As Dave and the others reached the decontamination line, they stared dumfounded for a few seconds in sheer disbelief and fright before they could react as the scene unfolded. Two of outside contractors had gotten into the jon boat, maneuvered across the lagoon, grabbed the gunnels tightly, and then sat back. Just then, on the opposite side of the lagoon, one of the two other contractor personnel got into the pickup truck and revved up the engine.

The second contractor grabbed the rope, and as the pickup engine would rev, he would loop the rope around the exposed wheel rim, pulling the jon boat forward a distance across the lagoon. They did this action several times. Looking like an old fashioned "Nantucket sleigh ride," and without life preservers, these outside contractors were going through this sleigh-like action of revving the engine, wrapping the rope around the wheel, and jerkily moving across the lagoon. It took about a half hour after first observing this crazy, unsafe, and unwise operation for the EPA/E&E crew to put a halt to it after finally convincing the contractor to stop or get thrown off the job. These were crazy early days in the industry, and many a cowboy took unnecessary risks in those early days.

Chapter 6

NICK AND THE SPY

If not a team of geniuses, they certainly were an assembly of youthful intellectual intensity—a mixture of highly educated young people all amassed at the same time in the same place at the same company. Scientists, engineers, planners, and skilled technicians of all sorts—marine, terrestrial, and aquatic biologists, archaeologists; geologists; civil and environmental engineers; planners—bugs and bunny types of all sorts. They were mixed with hard scientists, some of whom had backgrounds in industrial health and statistics, modeling, etc. They were highly motivated young and vibrant people brought together, forming a collective soup of intellectual egos.

Some partied hard while others had no social skills, but they all worked hard. Nick was sent a funny photo and caption sent by one of the field crew from a regional office while doing an assessment on a site in a rural area. The photo showed a couple of field crew members wearing Level C personal protection in a field with two wild turkey. The caption below the photo read "how do you expect me to fly with the eagles when I have to work with turkeys?"

Not long removed from his two-year raduate school stint in New Haven, Connecticut where he lived within the urban landscape a few blocks down from Saint Raphael Hospital along Chapel Street, Nick fell into this mixture and became part of it. He had taken a job with E&E in Buffalo, New York. Strange how college and grad school does not really prepare you for work in the corporate world; certainly not of this world. The elitist professors are lost in the acidemia and long ago forgot or possibly never understood the role of college for the majority of students. At least in the 1970s in hard science curricula, the focus of teaching was not on its real-life use and application of the subject matter. No, it was more on the subject's theory. Even so, it was far better than today where the focus that has crept into teaching is political wokeness and off-beat theory. It would not be the first or last time that Nick would learn to swim in unfamiliar waters using his guile and never fail attitude to get along despite his naivete.

Less than a year after graduation, Nick McCarthy sat in a small conference room at E&E headquarters office at Sugg Road in Cheektowaga, New York, with four other "older" guys that were flown in from various regional offices. These were some of the most knowledgeable and experienced people in the country in hazardous waste and chemical spill response. That's how young this industry was. There were not the legions of experts that would develop in ensuing years. These were literally some of the most knowledgeable people in the country when it came to hazardous waste investigation, spill response, and cleanup. E&E had assembled some of the most knowledgeable people around the country and stuffed them in offices in every EPA regional district. They mostly came from the Army Chemical Corp, DOD installations, and Coast Guard units or Navy nuclear submarine duty.

Their current mission, in this small conference room next to the E&E spill response nerve center, was to develop procedures for hazardous waste operations, investigations, and oil and chemical spill response. The first assignments under the freshly signed emergency response Technical Assistance Team (TAT) and hazardous waste site Field Investigation Team (FIT) EPA contracts was to develop standard operating procedures (SOPs) that would guide how work should be done. Still wet behind the ears and in awe of his colleagues—these tough-talking self-assured ex-military types—Nick just sat quiet, kept his mouth shut, observed, and took it all in like a sponge. He told himself to sit quietly in the corner and take lots of notes because his next assignment was to buy all the equipment to outfit every regional office and hazardous waste response van.

One of the very first procedures written was identifying and developing the various levels of personnel protective equipment. To date, there were no common levels of protection, because every military group/industrial location that delt with chemicals had a hodgepodge of haphazardly written rules—if any at all—to follow. It was this team's job to organize these into a set of EPA "Levels of Personnel Protective Clothing." Other procedures would also be written that provided guidance on downrange investigation procedures and pre- and post-medical testing requirements, equipment usage, and so on. As most of these individuals were ex-military of some sort, the procedures were full of military lingo. In this way, terms like command post and decontamination zone became prominent and some still exist in training manuals and the field lexicon today. Twenty-six-year-old Nick had been out from graduate school about a year when he was hired by E&E to work directly for the E&E president on his "special projects." Two of his first assignments involved writing white papers for use in grant applications and proposals—one regarding the environmental and human health effects of using tar sands and shale oil as alternate energy sources. Another assignment involved researching the potential effects of air pollution on the increased occurrence of otitis media—a disease of the inner ear that was affecting children at an increased rate.

Otitis media is inflammation or infection located in the middle ear. and can occur as a result of a cold, sore throat, or respiratory infection. Middle ear infections are usually a result of a malfunction of the eustachian tube, a canal that links the middle ear with the throat area. The eustachian tube helps to equalize the pressure between the outer ear and the middle ear. When this tube is not working properly, it prevents normal drainage of fluid from the middle ear, causing a buildup of fluid behind the eardrum. When this fluid cannot drain, it allows for the growth of bacteria and viruses in the ear that can lead to acute otitis media. The E&E president was convinced that the rise in otitis media was due to increased air pollution.

One of Nick's early jobs at E&E after the contracts had started in earnest was to manage the United States Department of Environmental Protection (USEPA) special project fund assigned to these FIT and TAT contracts. This was a separate budget, part of the contract E&E had with the USEPA, set aside as an emergency fund for all the regional offices. Whenever a regional office manager had the need for funds outside of their normal budget, they would call up Nick in the main E&E office in Buffalo and request an amount specific to a particular need. It was Nick's job to convince his counterpart at the EPA of the need and then document the transaction and deduct the amount from the special project fund.

One of the regional mangers was Edward Lee Howard, who was in charge of the Chicago E&E office. Nick liked Howard even though he had only briefly met him on a recent trip to Chicago. Nick mostly had communicated via phone to all the FIT and TAT managers. Howard came across to Nick as just a friendly, outgoing, and very competent voice compared to many of the others. He stood out and was easy to deal with. Nick was looking forward to meeting Howard and the other E&E regional managers at an upcoming meeting in Washington. Prior to being transferred to the Buffalo office to run the Corporate Health & Safety Department, Dave Dahlstrom ran the TAT-ERT (Emergency Response Team) office in Cincinnati. Dave later told Nick that Ed Howard visited his office on several occasions. Dave had two employees who were graduates of Russian Universities.

Edward Lee Howard had a master's degree in business administration from the American University in Washington. After his graduation, he worked for the United States Agency for International Development (USAID), which is the principal U.S. agency to extend assistance to countries recovering from disaster, trying to escape poverty, and engaging in democratic reforms. Ed was married to Mary Cedarleaf, whom he met in the Peace Corps in 1973 when they were both assigned to Bucaramanga, a town in Colombia. In February 1977, the Howards left for two years in Lima, Peru, where he worked as a loan officer US AID. Reportedly, the CIA sometimes uses US AID as diplomatic cover, but there is no information indicating that Howard worked for the CIA at that time. When they left Peru, Howard started his job at E&E in Chicago.

As Nick got to know his colleagues in the regional offices, he had no indication that Howard was unhappy. He was always engaging and pleasant on the phone but apparently his dream job was working for the CIA. In his job less than a year, the 28-year-old Howard applied to the CIA in 1980. It's unknown if there were ulterior motives then or later. However, Nick recalls very clearly a very strange, albeit brief, occurrence during their meeting in Washington. It was seemingly inconsequential, but for some reason very memorable.

Nick flew out of Buffalo to attend the all-hands meeting at the E&E Washington office located in the Washington, DC, suburb of Arlington, Virginia. There he met with the E&E staff that managed the EPA contracts and all the regional managers. It was a two-day meeting. The best part about the meeting was the chance to interact with and get to know all the regional managers, most of whom he only knew from phone or brief encounters during his health and safety trips to regional offices. The first day of meetings went great, and Nick was especially happy to meet Ed Howard, whom he had developed such a great relationship with over the phone the last six months or so.

They had a nice conversation during one of the breaks in the meetings, and Nick told Ed he was looking forward to their continued collaborations. The next day, they had a series of meetings during which Ed seemed actively involved. They all broke for lunch and headed out to the airport to catch their respective flights. A few of the E&E folks asked Nick to go with them to a restaurant that had great food but required a four-mile subway trip from their Arlington, Virginia, meeting to downtown Washington. As they were sitting at the table waiting for their food, Nick gazed across the crowded restaurant and to his surprise noticed Ed sitting a few tables away talking intensely to two individuals whom Nick did not know; neither was at the meetings or to Nicks knowledge E&E employees. Nick decided to go over and say goodbye to Ed one last time.

As he approached the table and said hi, Ed became very nervous and basically acted like he did not know who Nick was when just hours ago he was amiable and talkative. Nick quickly left, figuring he had interrupted Ed in some serious discussion and was obviously not welcome. It was the kind of strange experience—both awkward and weirdly unsettling—that stays with you, and you can recall it many years later. Before Nick never had an opportunity to talk with Ed again, he heard that Ed had resigned and was going to work for the CIA.

In January 1981, the CIA hired Edward Howard as a career trainee in the Directorate of Operations, also known as the DDO and as the Clandestine Services. Core collectors (CIA agents) must complete an 18-month Clandestine Service Trainee program, which trains and certifies them to handle tasks vital to foreign intelligence. Upon completion of the training program, new CIA agents are placed on two career tracks: Operations Officers (OOs) or Collection Management Officers (CMOs). As Ed spent several months at the secret CIA installation at Camp Peary, Virginia, near Williamsburg, he learned to be a spy. Edward Howard and his wife, Mary, were both employed by the CIA's Directorate of Operations, the agency's clandestine arm. They were trained by the agency to operate in Moscow as a husband-and-wife spy team.

Years later, rumors spread around E&E about the spy, the former E&E employee turned CIA agent who defected to the Soviet Union in the mid-1980s after a disappearing act in the New Mexican desert. Employees knew something was up when FBI agents showed up and disappeared behind closed doors with the E&E upper management. There were always rumors floating around the halls that E&E had some spies working for them using environmental consulting as a cover. It was not a real stretch to believe because E&E had been successful in some major federal contracts and had a number of former diplomats working for them, supposedly for their connections in Washington.

Ed Howard had been forced to resign from the CIA in 1983 after failing a polygraph test. At the time, he had been in training to operate in Moscow as a team with his wife. According to CIA public reports, in the spring of 1983, he had been getting ready for assignment to the CIA's Moscow station for his first overseas post, when at the last moment some troubling polygraph results and a security investigation disclosed drug use and petty theft. Instead of sending the officer to Moscow, the agency fired him.

The American authorities put him under surveillance after receiving information from Vitaly Yurchenko, a KGB deputy chief who defected to the United States in 1985, that appeared to incriminate Ed Howard. In one of the more famous cloak and dagger real-life spy stories, Howard, a CIA operative from 1981 to 1983, slipped away from FBI surveillance agents watching his New Mexico home in September 1985. Several months later, he was granted political asylum in Moscow. Trained as a Moscow case officer, he is the first CIA officer ever to defect to the KGB, and U.S. officials have said his disclosures caused the deaths of several agents working secretly in Moscow.

To the CIA, Howard had apparently looked like an ideal recruit. He had a graduate degree, work experience, and both he and his wife were accustomed to living overseas. Howard, fluent in Spanish and German, was a smooth, well-spoken man who collected guns and knew how to use them. Although born in New Mexico, he had grown up in Europe. His father, Kenneth Howard, had been an Air Force electronics specialist who worked on guided missiles and had been stationed at bases in Germany, Texas, and England.

While at the CIA, Howard was given five aliases and learned how to detect and evade surveillance. By June 1983, he was out of a job. He was now walking around with detailed knowledge of the agency's most sensitive operations in Moscow in his head and was reportedly furious at the CIA. Or perhaps this all occurred years before. Was the fact that he met with Russian University students during his visit to E&E's Cincinnati office or his peculiar reaction to Nick when he approached him in that restaurant in Washington just a coincidence?

A New York Times article in July 2002 reported the following: "Edward Lee Howard, the former C.I.A. agent who defected to the Soviet Union in the mid-1980's after a disappearing act in the New Mexico desert, has died. He was 50. 'The embassy has received reports that Edward Lee Howard died on July 12th,' said Richard A. Boucher, the State Department spokesman, confirming his death. Another official said the department had confirmed the death with Mr. Howard's next of kin.

"Mr. Howard's death remains as mysterious as his life. The Washington Post said he died of a broken neck in an accident at his dacha. But in a report by RIA-Novosti, the Russian government's news service, an unnamed Russian foreign intelligence officer, who said he knew Mr. Howard, denied 'this version of Howard's death,' but gave no further details. 'He is indeed dead,' Vladimir A. Kryuchkov, the former K.G.B. chief, said in a phone interview. Mr. Kryuchkov, who was imprisoned as one of the leaders of a coup against Mikhail S. Gorbachev in 1991 but who is now free, said he had received a call informing him of the death, but refused to give more information."

Peter J. Gorton

Chapter 7

Two stes in O Hi O

Prior to fully researching the various hazardous waste sites that amassed most of Nick's young life in 1983, I decided to take one of my many kayak trips. As I was rushing around to make sure I had my gear, I heard the telltale sound on my cell phone informing me I had a text. Not to be defrayed from my mission of not forgetting anything, I finished gathering my stuff—paddles, life vest, wet bag, water bottles, and a snack in case I got a low blood sugar attack in the middle. I double checked the gear and put it all in my truck and then looked at my phone.

The text was from my friend Willetta. "Hey Jed"—she called me Jed because she did not want to type out my full name Jedidiah. She asked me what I was doing in text lingo—"WYD"—then asked, "Want to go out tonight?" I answered sure and told her I'd check back later in the afternoon after my kayak trip.

My goal was to take a long kayak trip at one of my favorite places in the Erie Canal/Tonawanda Creek area. My plan was to put in in the Town of Amherst at the Veterans Canal Park and paddle due east along the canal/creek a little over three miles to Ransom Creek. I knew this was going to be a little longer of a trip than usual and figured my old body would start to yell at me before I made the turn back to the launch. But this was a trip I had wanted to make for a while. I had scouted out Ransom Creek from the road. Amherst Veterans Canal Park is located on the Erie Canal Bike Trail in Amherst, New York, and it's a great little park with a boat launch and kayak launch. Most of the time it's very quiet and not crowded except when a University at Buffalo event is going on.

The University at Buffalo Rowing group maintains a boat garage in the park. The park is accessed from Bear Ridge Road and Brenon Road at GPS coordinates 43.065077, -78.802716. Tonawanda Creek Road is directly across the canal from the launch, and cars go speeding by. The canal meanders along the road here and there, crossing over at various locations only to parallel the water on the opposite side. The canal is only about 200 feet wide in this portion give or take with some places wider and some narrower.

The water was like glass as it typically is unless it's a really windy day. Even then, the choppiness is minimal, making this a nice paddle for both beginners and mature kayakers. Once I got on the water, I headed upstream at a good clip initially wanting to get my usual workout in. Then I suddenly realized I was going much farther than normal and needed to pace myself for this longer than usual trip. I could have just put in at Ransom Creek, but there was no kayak launch there, and because I was getting older, it was a struggle to get in and out of the water without flipping the kayak and getting soaked. Being very spoiled now by these fabulous kayak launches, I decided I would just make the trip up the canal from the nearest launch, which turned out to be at Veterans Canal Park.

For those who may not be familiar with the Erie Canal, according to encyclopedia information, it is mostly a manmade water body that traverses east–west through western and eastern New York as part of the cross-state route of the New York State Canal System (formerly known as the New York State Barge Canal). When possible, the canal incorporated portions of existing waterways along the route.

It was completed in 1825 to create a navigable water route from the Atlantic Ocean with its terminus in Buffalo, New York, on Lake Erie and the Great Lakes Basin. When completed, the 363-mile canal connected the Hudson River in Albany to Lake Erie in Buffalo. This connection to the Hudson River allowed access south to New York City. The canal greatly enhanced the development and economy of Buffalo and other towns and cities along its route because it opened up a direct trade route from the east coast to the Great Lakes and the Midwest.

This section of the canal, which was the main focus of my many kayak trips, ranged from Pendleton, New York, continuing west and southwest to the Cities of North Tonawanda and Tonawanda. This portion of the canal really was part of the original channel of Tonawanda Creek. In Tonawanda, the canal intersects Ellicott Creek and farther west with the Niagara River where it turns south toward Buffalo. The beautiful thing is I discovered different kayak launches all along this portion of the canal one to two miles apart, offering a variety of sections to explore, including Ellicott Creek and parts of the Niagara River.

It was during this trip from Veterans Canal Park east toward Ransom Creek in early summer that I encountered the same thrashing of large fish I witnessed in the shallows of Annsville Creek. It was the third week in June, and at first I was caught off guard with the thrashing along the muddy creek bank by what appeared to be fairly large fish. I knew it was common for some very large lake fish to travel up the canal, but then I remembered my experience at Annsville Creek. As I paddled upstream at a steady pace along the shoreline, I first heard the splashing.

Every time I got close to get a look at what was causing the thrashing, it stopped only to resume farther up or down the bank. Finally, after some thought and leaning on my background in various biological sciences, I realized it must be some mating ritual. Putting that to rest in my mind and realizing I would never get a close look at the fish—and figuring I'd look it up later—I resumed my steady pace, wanting to make Ransom Creek before my old body started to rebel.

Apparently, carp spawn in the spring and early summer in weedy, grassy, shallow areas. When I took the time to research this spawning activity later that night, I learned that the splashing and other physical activity is the female laying the eggs with three to four males rolling their sperm. The dance spreads the fertilized eggs until they adhere to the grass and reeds and other objects in the water. The sticky eggs are deposited over these submerged objects as the several male carp spawn with a single female, the commotion making a noisy, muddy sight.

As I resumed my trip, I looked up from the scene to watch the huge billowing cumulus clouds doing their own dance across bright blue skies reflecting on the water so as to produce a mirror image. It was all so intense and calming. I realized how wonderful it was to stop my hectic life and allow nature to flow over my brain. By the time I made Ransom Creek, I realized the mistake I had made. My body was screaming at me, "You dope, you are not a young man anymore," as the result of 60-plus years of life and remnants of old sports injuries wreaking havoc on my lower back and base of my neck. I only went a few hundred yards up the mouth of the creek before I made myself turn around for the long paddle back.

Just then a swarm of very small fish passed by my boat in the crystal-clear water before disappearing instantly when they became aware of the ominous presence of the boat. The next day, I resumed my deep dive into the early years of Nick's career.

It was late February 1984, and Nick was sent to two sites in Ohio. The first site was the Chem-Dyne site in Hamilton, Ohio. Nick had flown into the Cincinnati

Airport on Wednesday, February 23, rented a car, and arrived onsite sometime around 10 am. It was a cold, sunny day with temperatures around 32 degrees Fahrenheit. Nick was just glad it was not December, because Christmas of 1983 was the coldest Christmas ever with blizzard conditions across northeast Ohio and northwest Pennsylvania into Buffalo. The combination of the arctic cold temperatures and the wide-open Great Lakes set the stage for the development of a lake effect blizzard. Snow fell across northern Ohio and northwest Pennsylvania on December 24th and 25th. Winds of 40 to 50 mph whipped between six and twenty inches of new snow into ten-to-fifteen-foot drifts.

Visibilities were zero for several hours, and virtually all roads were closed from just east of Cleveland, Ohio, to Buffalo, New York. Utilities were out for several hours, and hundreds of residents and travelers had to go to shelter areas. The City of Ashtabula was declared a disaster area, and the National Guard was activated. Several people suffered frostbite as air temperatures fell below zero. Some meteorologists called this the "Great Freeze of '83" and the worst cold weather event of the century in the United States. That would have made for terrible conditions for a site visit.

Now almost two months after the storm, Nick was sent to work with the E&E and EPA regional crew to complete an initial site survey and assessment of the Chem-Dyne property. This was one of the first times Nick had been to a site like this. It was a drum recycling site that had gone out of business, leaving hundreds of drums stacked two to three high across a field, leaking onto the soil forming a multi-colored scene.

Hamilton, located 20 miles north of Cincinnati, is a city in Butler County. Hamilton is the second largest city in the greater Cincinnati area and the tenth largest city in Ohio. By the mid-19th century, Hamilton was a significant manufacturing city focused on machines and equipment associated with the area's agriculture, including steam engines, hay cutters, reapers, and threshers. Other production included machine tools, house hardware, saws for mills, paper, paper-making machinery, carriages, guns, whiskey, beer, and woolen goods.

By the early 20th century, it became a manufacturing center for vaults and safes, machine tools, cans for vegetables, paper, paper-making machinery, locomotives, frogs and switches for railroads, steam engines, diesel engines, foundry products, printing presses, and automobile parts. During the two world wars, its factories manufactured products for war.

The EPA preliminary information and overview of the Chem-Dyne Site provided the following important geographic information: "Chem-Dyne is located at 500 Joe Nuxhall Boulevard (formerly Ford Boulevard) and occupies approximately 21 acres in the northern section of Hamilton, Ohio. The Site is bounded immediately to the south by a residential area and athletic fields. To the east are additional athletic fields and, beyond the athletic fields, a residential area.

The northern portion of the Site is bounded by the Ford Hydraulic Canal, which flows west to the Great Miami River; immediately north of the canal is a farm field. The Site is bordered on the west by railroad tracks. Adjacent to the railroad tracks is the Ransohoff Company, a sheet metal fabrication plant. Further west are the City of Hamilton Power Plant, warehouses for the Champion Paper Company, a small residential area, and the Great Miami River. Ground-water flow at the Site is generally west, toward the Great Miami River. Upon reaching the river, ground-water flow takes a more southerly turn.

The City of Hamilton has five production wells in its North well field, located 4500 feet north (upgradient) of the Site. The South Hamilton well field is located approximately 5 miles south of the Site, on the east side of the Great Miami River, and would be in the southerly path of the contaminant plume migration, were not for the hydraulic capture of the plume before it reached the water wells.

Production groundwater wells for Champion Paper Company, Mercy Hospital, Beckett Paper Company and Hamilton Electric Power are all located west and southwest (downgradient) of the Site, within the remediation area. No private wells were believed to be located such that they may be impacted by contamination from the Site."

The EPA summary also provided the following operational history of the Chem-Dyne Site: "The Chem-Dyne Site was operated as an industrial waste storage, disposal, and recycling facility from 1974 to 1980. During the years of operation, the Site accepted an estimated 112,000 drums of waste from approximately 200 generators. Materials handled included pesticides and pesticide residues, chlorinated and un-chlorinated solvents, waste oils, plastics and resins, PCBs, acids and caustics, metal and cyanide sludges, and laboratory wastes. Over 30,000 drums and 300,000 gallons of bulk materials were on-Site when operations ended in 1980. Most of the materials in 1980 were removed under the supervision of a state court appointed receiver during 1980-1981." After Nick's first visit and during his second visit later in the year, the remaining wastes were removed during a surface clean-up under EPA removal authority during 1983."

The site was placed on the National Priorities List on September 8, 1983. This trip to the site by Nick and the regional staff contributed to the follow up work just after the site's inclusion on the NPL. As mentioned, it would not be Nick's last time at Chem-Dyne.

Historical information indicates that chemical wastes may have been trucked to the site beginning in 1974. In 1975, Spray-Dyne produced antifreeze onsite from recycled chemical wastes. The operation expanded in 1976, and the Chem-Dyne Corporation formed. Wastes unsuitable for recycling were stored in drums and tanks or shipped to other disposal sites. Several environmental incidents were reported at the site, including fish kills in the Great Miami River and onsite fires from 1976 to 1979.

Nick left Chem-Dyne late Thursday night and traveled 100 miles northeast and arrived in Circleville, Ohio, which is just south of Columbus. He checked into a hotel, got a quick bite at the hotel restaurant, and went to bed for an early Friday morning site visit at Bowers Landfill. That next morning it was a balmy 43 degrees, and Nick's mission was to be the Site Safety Officer for a quick one-day site assessment in Level C personnel protective equipment. He would use an HnU Photoionization total organic vapor meter to monitor air levels as was required by E&E operating procedures for Level C protection.

Circleville is a city in Pickaway County and is set along the Scioto River about 25 miles south of Columbus. The city's name is derived from its original layout created in 1810 with the county courthouse built in the center of the innermost circle. Later residents changed the layout to the standard grid pattern that exists today. Circleville is home to the largest DuPont chemical plant in Ohio. It was opened in the 1950s. Truck manufacturing, tissue paper manufacturers, and vehicle transmission parts manufacturing are some of its other industries.

Bowers Landfill, also known as Island Road Landfill, was a formerly privately owned landfill site covering 12 acres near Circleville, on the Scioto River floodplain. The site operated between 1958 and 1968. Initially only domestic refuse was accepted, but from 1963 the site began accepting chemical waste from DuPont and other industries. Wastes were reportedly either dumped on the ground and covered with a layer of soil or burned in open fires. Prior to Nick's visit, EPA sampling of site surface water in 1980 showed the presence of contaminants, and in 1983 after Nick's visit, the site was added to the National Priority List (NPL) of hazardous waste sites. Long after Nick's initial visit and after numerous in-depth investigations, cleanup was completed in 1993, and the site was removed from the NPL in 1997.

Prior to Nick's visit, he read the following history of the site prepared by the EPA based on their initial work: "The land that became the landfill was purchased by John M. Bowers, a local dental surgeon, in 1957 and he began a sand and gravel quarrying operation on the eastern edge of the land adjacent to Island Road. Shortly afterwards, the portion of land between the quarry and the Scioto River was utilized as a landfill, with soil from the quarry used to cover the refuse."

During that time, the EPA listed sites such as Bowers Landfill on the NPL because they pose or had once posed a potential significant risk to human health and/or the environment due to contamination by one or more hazardous wastes. The once gravel quarry later became a residential landfill and then sometime after 1968, unauthorized dumping of chemical and industrial wastes, appliances, and used tires reportedly took place at the site. Because of its geologic sand and gravel nature, contamination rapidly migrated to groundwater and the nearby Scioto River. The eventual cleanup activities included removal of contaminated waste, landfill capping and venting, and revegetating the landfill surface.

Nick learned early during his studies in school and later during his many observations in their real-life application about the huge role wetlands play in nature. Wetlands are nature's way of cleaning the environment. It became clear to Nick over his lifetime as he saw the destruction of wetlands across America and in his own town outside of Buffalo how their importance in the environmental ecosystem was so undervalued. Yet another example for Nick was how the 7-acre wetland area in the Scioto River floodplain helped reduced contamination by absorbing impacted waters and slowly releasing them into the environment. At the site, later, wetlands were also created along the Scioto River, which transformed the area into a haven for plants, birds, fish, and animals.

Chapter 8

INDIANAPOLIS TO ASHTABULA, OHIO, AND A VISIT BY JAFOS

It was sometime around 1967, and twelve-year-old Nick walked into the woods with the gang: Billy Baker, Kevin Hyslop, Craig Cowan, Tommy Leaden, and Steve Sheridan. They were down on the lower part of the aqueduct near Oregon Road. They spent a lot of time up and down the aqueduct as kids using it as a conduit to other neighborhoods or to the woods where they built their forts. Every mile or so along the aqueduct there would be a stub road that went off perpendicularly and just ended a few yards into the woods. Nick didn't recall ever going down this stub before nor did he recall if it was Kevin, Steve, or Tommy who led the way that day. They had discovered something they wanted to show the rest of the gang.

It was during one of the early site visits around 1983 at one of the drum sites when Nick recalled this seemingly obscure memory from his youth. Nick suddenly relived in his mind the scene now so many years later. When the gang walked down that grass-covered stub road with two parallel areas long ago worn by tires from some vehicle, they walked into a clearing full off 55-gallon drums.

Nick was recalling the scene vividly now after all these years. They approached the drums, but something or someone in the group made them stop before they got too close. Now after all these years Nick knew exactly what those mysterious drums were. He had seen the same scene many times during his work career: a pile of drums randomly sitting in some field—just sitting there with the forest growing around them. He knew what it was. Some unscrupulous people who did not want to pay the cost of proper disposal had dropped these off in the woods, probably in the dark of night. Hazardous waste drums dropped off by someone who drove off Oregon Road and up the aqueduct about a 100 yards, disposing these drums in the woods. Luckily, the boys never got close enough to be exposed or, worse, tip them over as boys are wanting to do. Something stopped them, and they left the area with the memory stocked away in Nick's mind. It was just a momentary event, but it obviously made an impression.

Nick left very early in the morning on a day in early May 1983 to get to the Buffalo Niagara International Airport. He was embarking on a somewhat dizzying seven- or eight-day trip starting in Indianapolis, Indiana, and ending across the State of Ohio in Ashtabula. He was chosen to go on this trip to visit potential Superfund sites because he was one of the more experienced field members of the Buffalo Office Hazardous Waste Team and had proven his merit by getting excellent work done when sent out alone or with a team. Nick and another from a different office were to visit and assess four separate potential Superfund sites and complete reports to meet a USEPA milestone.

The EPA at the time called their task a RAMP walkover. RAMP stood for Remedial Action Master Plan. They were to gather the initial information for later use in the understanding of and final environmental remediation of these hazardous waste sites. Essentially, they were some of the first to visit these sites and were walking into somewhat unknown hazardous conditions. He was to arrive at Indianapolis, rent a panel van to store all the equipment he brought, including real-time organic vapor meters, and meet Greg who worked for CH2M Hill in their Seattle office. CH2M Hill was a partner with Nick's firm Ecology and Environment (E&E) on this first ever Field Investigation Team (FIT) contract. Greg was coming from the Seattle office, and they had never met aside from speaking on the phone. Nick could tell from the few conversations they'd had that he would probably like this guy.

"You Greg?" Nick asked as he saw this late-20s/early-30s-year-old guy wander toward him at their pre-agreed meeting place. "Yeah," Greg said with a smile as they reached to shake hands. You know what they say, you can usually tell a lot about a person the first few moments you meet them. Greg, despite being a technical person, had the appearance and gestures of a "normal guy." The two men loaded the van with the luggage and supplies Greg brought and proceeded northwest on U.S. Highway 421 to their first scheduled destination; the Envirochem hazardous waste site in Zionsville, Indiana. Greg had remarked to Nick as they started their journey together, "This is the farthest east I have ever been. Everything in the Seattle area is so new; roads, buildings, landscape. Everything here seems so much older with such an older history."
Nick, who had been to both Seattle and Portland, Oregon, as well as to Kent, Washington, replied, "Yeah, that is interesting. And I bet even the terrain looks

so different. I recall how struck I was with how lush and green everything was in Seattle and Portland and Washington the first time I went there." This was the start of many interesting conversations the two had as they became fast friends driving together from Indianapolis across Indiana and Ohio. Nick recalls thinking, "I'm so glad I am with someone I get along with as this would be a long week if I had to do this intense schedule with a person I did not like." It was a lesson he would learn the hard way a year later when he would spend a month rooming with someone he did not get along with on his mission to Johnston Island.

Envirochem was located in Zionsville, which is a suburban town located in the extreme southeast area of Boone County, Indiana, just northwest of Indianapolis. They arrived in late morning on May 9, 1983. According to the background information supplied to Nick and Greg, Envirochem was first acknowledged as a hazardous waste site in April 1979. It would later be listed as a National Priority List site in September 1983 based partially on the information collected by Nick and Greg during this trip. This would not be the last time Nick would be on this site. He spent two more memorable occasions on this particular landfill creating memories that became etched in his mind.

Envirochem, also known as the "Environmental Conservation and Chemical Corporation" or the "ECC" site, started operations in 1977. Like many sites Nick had been on during the early years, this is what Nick referred to as a "waste and drum recycler site." This site took in containers of chemical wastes for recovery, reclamation, and brokering. These included solvents, oils, and other wastes received from industrial clients, which arrived in drums and bulk tankers. Nick would come to know these sites well as his career moved along. They all suffered from poor management and maintenance and badly run operations. The site was adjacent to the Northside Landfill. The facility recovered solvents and oils from industrial sources from 1977 until May 1982, when it was closed under a court order obtained by the state.

When Nick and Greg arrived onsite, they were amazed. They had never seen a site quite like this! Drums were stacked haphazardly two to three high, and a chemical stew was visible on the ground as leaks from many tanks and drums left a colorful rainbow of puddles impacting the soil in various areas across this six-acre site. Later, Nick found out that over 20,000 drums and 400,000 gallons of waste remained onsite.

As Nick and Greg first conducted a perimeter review, they noticed that the Northside Sanitary Landfill (eventual Superfund site) was located immediately to the east of the site and that another impacted site—later named the Third Site was immediately to the south. They came upon an unnamed ditch near the west side of the site and observed that it flowed into Finley Creek, which flows into Eagle Creek and eventually into the Eagle Creek Reservoir about ten miles farther downstream.

Contaminated rainwater from a holding pond flowed into this unnamed ditch, causing an oily sheen on Findley and Eagle Creeks. Organic solvents such as 1,1-dichloroethane, trichloroethene, and 1,1,1- trichloroethane were later found in onsite wells. The surface runoff impacted nearby surface waterbodies and sewers and the inevitable migration through soils eventually impacted groundwater. Only the geology, time, and a few other physical and chemical parameters would determine the extent of the impact. Typical of these sites, Envirochem became insolvent when the cost of even basic proper environmental requirements was too much for the ill-managed operation, and it was placed into receivership in May 1982. It would take many years after Nick's initial visit to remediate the site with what was hoped to be the final remediation completed in 1998.

The next time Nick was at the property was about a month later, just prior to the first serious assessment and investigation of the site. Nick flew out to the property a day before he was to meet several higher-ups in CH2M Hill and E&E. His job was to function as health and safety officer while he took the gentlemen for a quick tour.

Nick spent the first day before the entourage arrived reacquainting himself with the site. That's when he noticed it. He had walked down range scoping out the route for the next day's excursion when it happened. The noise scared him instantly, stopping him immediately in his tracks. He had not planned on venturing very far because he was by himself and that was certainly not protocol on a site of this nature. Looking around trying to gauge what caused the noise, it happened again. This time he saw it too. Walking back to the site trailer, he asked the onsite coordinator about it and was given the explanation.

The next day as he rounded up the group and got them dressed in Level C protection for a quick site tour, he first gave a site safety briefing that included a brief history of the site and concerns specific to their site walk. He left out the one key observation he had made the day before, figuring should it happen again it would make for a teachable moment and be a good way to get the attention of these higher-ups who rarely if ever had been on any site and certainly never one of this magnitude. As they started down through an isle of drums, it happened again—then again and again in quick succession. It definitely caught the attention of the group; they all stopped in their tracks.

It was late-morning, and the early summer sun was raising the ambient temperature. As it did so, the chemicals in some of the drums started to expand. First one drum lid popped as it expanded outward, and the force made the drum move slightly or "jump." Then another drum jumped until a few random drums in various stacked piles popped and moved—almost dancing to the ominous sounds. Well, you can just imagine the expressions on the faces of the visitors.

Nick smiled as he informed the visitors what was happening. One asked, "Should we be here? Will they explode?" Nick answered, "We certainly can leave, but just remember these drums have been here for many years now, and they likely have done this many times. Besides," he assured them, "we'll be gone long before it gets really hot!" A short time later he thought better realizing how you never know when the day may be different and the reaction more severe, so he cut the visit short realizing they had seen enough. He reported these events to his superiors and team members back in Buffalo concerned perhaps some action may be needed.

Based on the initial work that Nick and Greg did, CH2M Hill performed a Remedial Investigation (RI) in 1983 and 1984, which involved an investigation of the nature and extent of contamination in the soil, groundwater, surface water, and sediments on and around the site. Soil contaminants were found to contain volatile organic compounds (VOCs) and phthalates. Migration of VOCs in the soil to the shallow saturated silty clay and shallow groundwater zone had occurred on the property. Organic contaminants were also found in Finley Creek immediately downstream of the site.

The last time Nick was onsite was during the middle of this CH2M Hill Investigation. He arrived onsite and immediately went to the onsite trailer/command center and reported to the U.S. Army Corps of Engineers project manager (PM). The drums were all gone, and Nick's job was to sample the near surface soil for contamination to determine where initial clean-up should occur and how much soil should be removed. The EPA had completed a limited feasibility study, and disposal of all drum and tank wastes and onsite treatment and disposal of contaminated rainwater had taken place. Due to the deteriorating conditions at the site, they took emergency measures to stabilize it. Concurrently, the USEPA/Corps of Engineers initiated a remedial investigation and feasibility study to determine the cost-effective remedy for dealing with contaminated soils and ground water.

At that time, the U.S. Army Corps of Engineers (USACE) had just started providing technical support to the EPA Superfund project, which was enacted by Congress in Public Law 96-510, the Comprehensive Environmental Response, Compensation, and Liability Act of 1980 (CERCLA). This legislation made provisions for the Superfund program's use of other federal agencies' existing capabilities in meeting its objectives. While Nick had the ultimate respect for the engineering merits of the Corp, he knew they were relatively new to the hazardous waste site work. In this case, the PM was relatively young and green.

Nick established his mission with the PM and collected his crew to create a grid pattern across the site where the drums had previously been located. But about a quarter of the way through the sampling, Nick noticed a contractors' bulldozer moving around the soil they had just sampled. Nick marched back into the trailer after exiting decon to speak to the young PM.

"Not for nothing," Nick said as he pointed out the trailer window at the bulldozer, "but the contractor is bulldozing around the soil we just sampled, making the samples we just took meaningless." Nick continued with the air of someone with experience: "I suggest you put a halt to that because he is basically spreading potentially contaminated soil across areas that may not be contaminated, and the soil samples we collected have no bearing on the soil anymore."

Nick continued, "Our sampling program is supposed to delineate the contamination, so you know the volume to remove!" The PM did not seem to understand Nick's complaint, which made Nick more agitated. More likely, he did not care. One thing Nick had noticed working for the government was that many times people just followed "orders" without any care about the reason. Nick followed with "Have you guys moved any soil out there after the drums were removed?"

The PM told him no. Now, Nick was very cognizant of costs and work effort because he had a good deal of experience at this point. He rapidly realized two pretty important factors: 1) the contractor, unknowingly or on purpose, was creating more soil requiring disposal at a significant cost, and 2) There was really no reason for doing this sampling in the first place since leaking drums had been stored in this area for years.

"Well," Nick went on, "the first time I was on this site there were stacks of drums and tanks at least two-three high leaking all sorts of stuff all over the ground; the ground was a chemical stew! My suggestion to save money is to remove a foot or two of the soil off that entire drum storage area and then come back to sample the soil below."

The young PM was getting a little angry at Nick's self-assured, demanding statements, and he reacted in a typical military-rank confrontational way. He looked at Nick and said, "Just go out and sample the soil like we hired you for." At that, realizing his company would get a bad mark if he continued, Nick turned around and completed the mission, leaving just another negative mark in his mind against bureaucracy, government spending, and waste. He reminded himself that it was a good thing he never served in the military because he knew he would never be able to take the mind-numbing order of command.

It was during these early days that Nick first realized the waste and abuse of government and how these sites seemed to be on a never-ending mode of study and investigation, costing significant amounts of money before they ever got to even the simplest overtly obvious long- term remediation. "Oh well, the source – the drums were gone and at least that was a huge accomplishment" he thought at he time.

The EPA official documentation provided the following summary of the site: "Enviro-Chem began operations in 1977 and was engaged in the recovery, reclamation, and brokering of primary solvents, oils, and other wastes received from industrial clients. Waste products were received in drums and bulk tankers and prepared for subsequent reclamation or disposal.

Accumulation of contaminated stormwater on-site, poor management of the drum inventory, and several spills caused State and U.S. EPA investigations of Enviro-Chem. The State pursued Enviro-Chem for violations of the Environmental Management Act, the Air Pollution Control Law, and the Stream Pollution Control Law, resulting in a July 1981, Consent Decree approved by the Boone County Circuit Court. That Court imposed a civil penalty against Enviro-

Chem and placed Enviro-Chem into receivership. In May 1982, Enviro-Chem was ordered by the court to close and environmentally secure the Site for failure to reduce hazardous waste inventories. By August 1982, Enviro-Chem was found to be insolvent. U.S. EPA placed the site on the NPL list in September 1983."

A Remedial Investigation (RI) was conducted in 1983 and 1984 (after Nick and Greg's initial assessment), which involved an investigation of the nature and extent of contamination in soil, groundwater, surface water and sediments on and around the Enviro-Chem Site. A Feasibility Study (FS) was completed in 1986, which evaluated several alternatives for cleaning-up the Envirochem site and the neighboring Northside Landfill site, which had also been placed on the NPL.

Surface contaminants were removed from the Enviro-Chem Site in an operation extending from March 1983 through 1984. These cleanup efforts were initiated by the EPA and completed by a group of potential responsible parties (PRPs), overseen by U.S. EPA and IDEM, pursuant to a Consent Decree entered on November 9, 1983. Actions included removal and treatment or disposal of cooling pond waters, approximately 30,000 drums of waste, 220,000 gallons of hazardous waste from tanks, 5,650 cubic yards of contaminated soil and cooling pond sludge.

In March 1985, contaminated water was discovered ponded on the concrete pad at the southern end of the Envirochem site. During the resulting emergency action, EPA constructed a sump at the southeast corner of the site and removed and disposed of 20,000 gallons of contaminated water containing high levels of volatile organics.

On the initial visit a year or two before Nick's confrontation with the Corps of Engineers young PM, Greg and Nick spent a day and a half on Envirochem before they headed east out of Indianapolis on Route 70 to their next site: Fultz Landfill in Cambridge, Ohio. They checked in at the Days Inn on the evening of May 10, 1983, after spending the night before in a Knights Inn. Both Nick and Greg found the Days and Knights Inn theme rather ironic as it was maintained on their trip as they traveled across Ohio on Route 70.

When they reached Cambridge, Ohio, approximately 75 miles east of Columbus, they went about one mile southeast on Interstate 77 to Fultz Landfill. The landfill was owned, developed, and operated by Mr. Foster Fultz from October 1954 until his death in June 1982. The 30-acre Fultz Landfill site was located in east-central Ohio in a rural agricultural area. The site was on Nick and Greg's list because it had been alleged that hazardous industrial wastes were disposed in the landfill. It was not until 1985, two years after Nick's visit, that the landfill was officially closed. The history indicates that prior to 1950 and before becoming a landfill, it was part of a large farm that comprised about 200 acres. The landfill was on a moderately sloped ridge that overlayed an abandoned subsurface coal mine in the Upper Freeport Coal seam, a portion of which was on the former strip mine drainage swale. The swale reportedly intermittently channeled storm water runoff into nearby Wills Creek. A number of other strip mine ponds existed along the drainage swale north of the landfill.

Nick would find himself on many such rural landfill sites throughout his career that were nothing more than an open dump typical of landfills before the 1970s and 1980s. The site was first licensed by the county to accept household, commercial, and industrial solid waste. However, after an accident with a drum and a resulting fire in 1983, just prior to Nick and Greg's visit, the site was reported to local and state authorities, and a landfill employee confirmed emptying industrial solvents in the landfill to recover the drums. The initial work by Nick and Greg led eventually to detailed investigation and remedial plan in June 1991.

On the evening of May 11, 1983, Nick and Greg left Fultz landfill and traveled along Route 70 to Buckeye Landfill in Saint Clairsville, Ohio. As they started on their trip, Greg asked, "Good lord, is there nothing but corn in Ohio?" As the two traveled along, there was what seemed to be an endless flat terrain of corn rows as far as the eye could see, mile after mile. However, when they were passing through more populous cities and towns, Greg again commented on the "old" look of everything, and Nick told him that on the east coast things were much older.

He told Greg about the Hoxie House in Sandwich, Massachusetts, on Cape Cod near his parents' home and how he recently visited the house and learned its history. Nick told Greg that the Hoxie House was built in 1675 by Reverend John Smith, who lived there with his wife, Susanna Hinckley, and their 13 children. Abraham Hoxie, a well-known whaling captain, purchased the home in the 1850s. Until the 1950s, the property was a family home without electricity, central heating, or plumbing.

The Hoxie House was never updated with modern conveniences and now is a museum designed to teach visitors about the simplicity of colonial life. Nick told

Greg how interesting it was to go through the house with a guide and learn all sorts of little trinkets of knowledge, like how the expression "sleep tight, don't let the bed bugs bite" came to be. Nick told him "It provides you a better understanding of what life was like in the past and how strong and resilient people were in those times with little government control and confinement. They were true rugged individuals."

A 1991 United States Environmental Protection Agency Ohio Environmental Protection Agency Plan Wrote the following about the Buckeye Landfill years after Nick and Greg's Visit In 1983: "The Buckeye Reclamation Landfill Site was located in Richland Township, approximately 4 miles southeast of St. Clairsville and 1.2 miles south of Interstate 70 in Belmont County, Ohio.

Approximately 200 homes were located within a 2-mile radius of the site, downstream of the site boundaries. Deep mining occurred beneath the 658-acre site until around 1950. During that time, the site was a disposal area for mine spoils which were disposed on the ridge west of King's Run and in the drainage ravine for King's Run. The site was licensed as a public sanitary landfill in 1971 and operated by Ohio Resources Corporation, under the name of Buckeye Reclamation Company."

About 50 acres in size, the facility accepted general trash, rubbish, and nonhazardous waste as well as industrial sludges and liquids from municipalities and villages, including asbestos, carbon black, and fly ash. Soon after Nick and Greg's visit, the Buckeye Reclamation Landfill was listed on the National Priorities List (NPL) by publication in the Federal Register on September 8, 1983.

Nick and Greg only spent a single rather long day at the Buckeye Landfill, and early the next morning they checked out of the Days Inn on their way to their next destination: the New Lyme Landfill in Ashtabula, Ohio. They made the two-and-a-half-hour trip due north to the shores of Lake Erie just outside the City of Ashtabula near the mouth of the Ashtabula River. History indicates that Ashtabula was a key destination on the underground railroad during that shameful period of America's past. Beginning in the late 19th century, the city evolved into a major coal port on the Lake northeast of Cleveland.

The New Lyme Landfill was located near State Route 11 on Dodgeville Road in Ashtabula County, approximately 20 miles south of the City of Ashtabula, Ohio.

The landfill occupies about 40 acres of a 100-acre tract. The site was operated by two farmers beginning in 1969. By 1971, the landfill was licensed by the State of Ohio, and operations were taken over by a licensed landfill operator. According to documentation, the New Lyme Landfill received household, industrial, commercial, and institutional wastes and construction and demolition debris. The Landfill was cited on numerous occasions for improper disposal practices and in August 1978 was closed by the Ashtabula County Health Department. Documents indicate that wastes at the New Lyme Landfill site included coal tar and coal tar distillates, asbestos, coal tar, resins and resin tar, paint sludge, oils, paint lacquer thinner, peroxide, corrosive liquids, acetone, xylene, toluene, kerosene, naptha, benzene, linseed oil, mineral oil, fuel oil, chlorinated solvents, 2,4-D, and laboratory chemicals.

Nick would visit Ashtabula on a number of occasions after this first visit with Greg as well as other nearby sites which includes the Old Mill site in Rock Creek, Ohio. Old Mill was registered as an active NPL superfund site by the EPA and was considered one of the worst hazardous waste sites identified by the EPA. Land use in the vicinity of the site was represented by a mixture of residential, agricultural, and commercial/light industrial developments. The closest residences were just across the street from the property boundary. In 1979, the EPA and Ohio EPA found approximately 1,200 drums of toxic waste, including solvents, oils, resins, and polychlorinated biphenyls stored on the property.

Prior to Nick's arrival, emergency removal activities had taken place that resulted in drum removal and at the Old Mill site excavation of contaminated soil from the drum storage areas. Nick's job was to sample soil at various locations across the site, which at that time was made up of a series of old, vacant agricultural looking buildings with dirt floors. The site was proposed for inclusion on the National Priorities List (NPL) and after Nick's visit was included on September 8, 1983.

During the site investigation, Nick's team collected surface soil samples, subsurface soil samples, and groundwater samples after installing a series of groundwater monitoring wells on the property. For this job, Nick worked with a two-man crew from an Ohio drilling firm that he had worked with previously on two or three other jobs, including the Ashtabula Landfill site. He had become friends with these two; they were the type of friends that are made working 10-hour days in hazardous environments in a high level of personnel protective clothing.

It was at this site that Nick had one of his more memorable and funny adventures. The first day onsite while the drilling crew was setting up for borings he went downrange with another E&E colleague to sample surface soils from some of the dirt floors where drums had been stored. For this task, the crew wore Level C protection, which included full-faced respirators and combination dust/organic vapor cartridges, poly-rubber suits, gloves, hoods, and boots. Nick was scrapping up soil to put in the sample jars his partner was holding. As he did, he noticed these elongated tootsie-roll-looking gobs of soil. Thinking they may be remnants of what was in the drums, he carefully filled two sample bottles, making sure he got these and other soil into the jars for later laboratory analysis. Halfway through the third jar, he burst out in laughter so hard he almost lost the seal on his mask.

He laughed even harder when he looked at his partner's vividly perplexed face, which was obvious even through the full-faced mask. Once he regained his composure after about two or three "What's so funny?" questions from his partner, he explained that he had just filled the sample jars with dog crap. It had dawned on him after he looked a little closer at these "rolls" just what they were. He had cleaned enough of them up in his backyard as a youth from his mom's two miniature schnauzers. Apparently, the little mutt from the neighbor's house that they saw hanging around the site was using this place as his preferred bathroom. The two laughed harder when they thought about the lab analyzing dog shit.

Since the jars had enough other soil in them, they decided to send them as they were. The next day, when they were walking downrange past this spot to start the drilling program, they became a little more serious. It had snowed just enough to cover the ground with a thin coating overnight, and the snow in the area they had sampled the day before was stained purple, pink, and red, indicating something was in the soil they were laughing about the day before.

The investigations found that the soils were contaminated with tnchloroethene (TCE), dichloroethene (DCE), 1,1-DCE, vinyl chloride, 1,1,1- trichloroethane, ethylbenzene, and xylene, with TCE as the principal contaminant of concern. The soils were also contaminated with heavy metals, such as lead. Groundwater was found to be contaminated with TCE and other organic chemicals.

There were many memorable times with these drillers, mostly silly things that men do when thrown together in hazardous work. Every morning they would meet for breakfast at the only restaurant in this small town. The first time they had breakfast at this place, one driller ordered pancakes. After a few bites, a funny look came across his face, and he motioned to ask the waitress what was in these cakes. "Just normal pancakes, she replied: 1½ cups all-purpose flour, 3½ teaspoons baking powder, and ¼ teaspoon salt." Why?" The driller replied without missing a beat, "Gee, these hot cakes are stomach busters. These are the densest pancakes I ever had. Did the cook make these out of bentonite?"

Bentonite is an absorbent swelling clay that the drillers use when constructing groundwater monitoring wells. It is used to seal off the well screen from groundwater seeping down from the surface. Once the bentonite pellets come in contact with water, they immediately swell, forming an impenetrable layer. He followed, "I took two bites and they swelled in my stomach in seconds. Ask the cook if we can take some of the batter with us for use on our wells. Holy cow, this is better than any bentonite I ever used!" Every day after that, he told the waitress he did not want any of those bentonite pancakes.

Then there was the other time at the same restaurant about a week or two into the job when the same driller asked the waitress who had been waiting on them every day where all the women were in this town. The waitress replied, "They all get married at 16," and without missing a beat the driller followed his question with "Where are all the 15-year old's?" as the table and waitress erupted in laughter. The little things are what the crew would find funny after monotonous days in the hazardous environment.

It was these two drillers who taught Nick one of his favorite expressions he would use the rest of his career. These guys were hard workers, but they were also cut-ups. One day, a week or two into the assignment, Nick told them that the EPA project manager and a few others were coming to visit the site. One of the drillers looked at Nick and said, "Great. More JAFOs." "JAFOs?" Nick repeated with a quizzical look. "Yeah," replied the driller. "Just Another Fucking Observer."

Nick used this expression often the first time he would meet a contractor crew that he had to oversee. Typically, he would wait for one to ask what his job was, and Nick would say without missing a beat: "I'm the JAFO on this job." He always got a laugh after explaining what a JAFO was, and it usually broke the ice and this typically got the contractors on his side from the start.

Chapter 9

.22-CALIBER KILLER AND THE MIDTOWN SLASHER

It was interesting to Nick to observe the little things around a complex event, which is in the nature of scientists. You know the expression "Don't sweat the small shit." Well in Nick's case, this was often hard not to do, and it often paid off. Was it part of his training, or is it the unique nature of people in that field? It's a sort of chicken and egg thing, but obviously the two both play a role. Some appear better at it than others; was it a natural gift provided them or just an enhanced desire to contemplate things more deeply?

This sometimes unfortunately included trivial things that may take away the time that should be spent on more important considerations. Or were they trivial? Take Nick's observation of how people operate in the field their first time. Nick spent lots of time at hazardous waste sites that were in commercial districts in towns and cities but also lots in rural areas where nature had reclaimed its territory, hiding the waste and other dangers in clever ways.

Since this was the dawn of this industry, it was the first time for most of the people Nick encountered. We all know about first times. They typically lead to some interesting stories wrapped around mistakes or luck—or the opposite. Many of the early hazardous waste workers were either ex-military workers in the hazard chemical and biological units who had gone on to get a degree or kids right out of college and university with no real-world experience. Nick often observed how easy it was to spot a "city" person who never had spent much time in the woods. It was typically clear on their first trek across a site that certain people just had no clue how to navigate walking in the woods, and it would ultimately end with them in a briar patch or walking through inch-or-more-deep muck in the lowest part of the property.

Now, Nick did not make these observations out of superiority. No, it was merely something that came to him as a scientific observation, and he would always respond by pointing out in a delicate non superior way the proper approach to enter and traverse a wooded area. This was something that was second nature to Nick from his years in the woods growing up when he and his friends Billy, Craig, Tommy, Steve, and Kevin would set off on some days-long adventure into the woods near their Town of Cortlandt homes just outside Peekskill, New York. He never made fun or lost patience with these people even if he made the mistake of following them into a dead-end only to have to circle back and take a better route. Only a few times did he raise his voice and point out how to "read" the topography or lay-of-the land in time to avert being led over the edge of a rock cliff. This was usually a teaching moment: Nick would remind them the importance of looking at the site maps and aerial photographs or just the way to see the patterns of the woods, the lay of the land, the "look of the greenery."

Nick had spent enough time in the cities to understand its patterns too. He had acquired the knowledge coupled with his innate ability to maneuver around that environment as he was sure these "city" people were also expert at: how to maneuver through massive numbers of people scurrying here and there and cars and dead-end alleys in dangerous parts of town, or the complexity of taking the subway in a large city—when to take the A train and what stop to get off to catch the B train.

Nick had been back in Buffalo for about a week from one of his trips to audit one of the regional offices and was heading to the Cheektowaga Sugg Road office early. He parked his car in the lot along the rail tracks not far from the only two other cars in the lot. As he stepped out of his vehicle, he noticed a large red stain that looked like blood leading to the brush along the edge of the lot and the adjacent rail tracks. Some of the shrubs and weeds appeared to be bent over as if something heavy was dragged across them towards the tracks. He noted this scene as he walked towards the office door, his mind swirling. He proceeded upstairs and into the large room that held the Drafting and Word Processing Departments. Back then, in the early 1980s, no one had personal computers on their desks. Figures needed for the technical reports were hand drawn and given to one of the draftsmen to create a formal figure, and reports were handwritten and given to the typing pool/Word Processing Department.

Nick walked into the area and was greeted by Pat, who was one of the draftspersons, and Don, who headed the department. Nick said to both as he approached: "Did you see all that blood in the parking lot?" Don looked at him solemnly and said, "Yeah, we just called the police. They are on the way." Nick looked at both of them and asked, "Do you know what it's from? What happened?" Pat was one of Nick's friends, and he often talked to her about work and the company and some of his field projects while explaining what the figure he was giving her needed to depict. Pat, visibly upset, replied, "We don't know, Nick, but it sure looks pretty bad. Don and I were just discussing that when you walked in." Nick replied, "Well, whatever happened, the thing that lost all that blood is not alive. That is an awful lot of blood." Nick looked at both of them as he turned around, and said, "I'm heading back outside. I want to meet the police when they arrive and show them what I saw." Pat replied, "Go ahead. I'm staying here because the whole thing is freaking me out." Don nodded in agreement as Nick was exiting through the door.

Just as Nick got back outside, he saw the cop cars motoring in at a good clip, and he motioned them over to the scene. Nick said a few words to one of two detectives who began to review the scene. As one police officer focused on the pool of dried blood, the other followed the disturbed foliage to the rail tracks. Nick asked the detective at the blood scene what he thought. "Probably someone gutted a deer," was the short answer that came back as Nick realized he needed to back away and leave them to their work. As he walked away, he saw something white a few feet away from the scene. As he got closer, he bent down and saw it was a tooth. The tooth had a filling. "Hey officer," Nick shouted to the policeman.

"Over here!" As the officer approached, he pointed to the tooth. "I don't think they are putting fillings in deer," Nick said as the officer bent over to look closer. At that, the officer looked much more serious now, and said, "Sir, I think you should go back inside. We'll come in and talk to you later."

Between September 22 and September 24, 1980, four black men were shot in the head at close range with a .22-caliber pistol in the Buffalo, New York, area. They were either walking along streets or sitting in cars. Then on October 8 and 9, 1980, two cab drivers were bludgeoned to death and their chests were slashed in a similar horizontal fashion with their hearts cleanly removed. Erie County District Attorney Edward Cosgrove called the mutilation murders "the most bizarre thing I've ever come across in my life." The two cab drivers, Parler

Edwards, 71, and Ernest Jones, 40, both of Buffalo, had their chests slashed in similar fashion. Edwards also was beaten about the head, and Jones's throat was slashed. Headlines screamed: "Police Search for a 'Deranged, Mentally Disturbed' Killer." There was never any evidence linking the cab drivers' deaths to the shooting deaths of four black men, but the head of an investigative effort that included federal, state, and local law enforcement agencies claimed the killings were the work of insane assailants. Additional assaults followed, including assaults in Manhattan, Buffalo, and Rochester, New York.

The police never came in to talk to Nick. Nick read in the paper the next day that Edwards was found stuffed in the trunk of his cab a few miles away from the office in Amherst, New York. The body of the other cab driver, Jones, was found near a harbor in the town of Tonawanda.

Joseph Gerard Christopher was a Buffalo native who grew up in a predominantly Italian neighborhood. He reportedly adored his father, who was an avid outdoorsman and hunter and taught his son how to shoot and handle weapons. His father died in 1976, leaving Christopher with his mother, a nurse in a nearby hospital, and three sisters. He attended parish classes at a Roman Catholic church but finished in public schools where he entered the automotive mechanics program. He eventually dropped out in 1974 and was remembered as a quiet, unassuming student.

In the late 1970s, he tried to enlist in the Army but was unsuccessful due to a medical condition. This was followed with a series of odd jobs. At one of his jobs, he developed a relationship with a gun instructor and obtained a pistol permit with the goal of becoming an instructor. After his relationship with the instructor ended, his life reportedly began to spiral out of control. He reportedly turned to the Buffalo Psychiatric Center for assistance, and rather than being admitted, he was recommended counseling therapy instead. Fourteen days after he left the center, the killings began.

On November 13, 1980, Christopher finally managed to enlist in the U.S. Army and was stationed in Fort Benning, Georgia. He was given a Christmas furlough and went to Manhattan, New York City, arriving there on December 20. Two days later, a murder spree attributed to the so-called "Midtown Slasher" began where five African American men and one Hispanic were attacked in stabbing attacks, over a twelve-and-a-half-hour span. Returning to Buffalo, Christopher reportedly knifed four more black men from December 29 until January 1, 1981. Later, back at Fort Benning, Christopher attacked a black soldier with a paring knife, and he was placed in the fort's stockade. He attempted suicide, and in a subsequent psychiatric session he admitted to having to" kill blacks." A subsequent search of his home found evidence linking Christopher to three of the murders. He was indicted in April 1981 and he was transfered back to Buffalo for his trial on May 8.

Known as the ".22-Caliber Killer" and the "Midtown Slasher," Christopher was an American schizophrenic serial/spree killer who reportedly murdered at least eleven African American men and one Hispanic over a five-month span, between 1980 and 1981. There is still some question as to whether he was responsible for the cab driver murders. During the murder spree, a famous FBI profiler, John Douglas, worked the case in Buffalo. Douglas was renowned for examining crime scenes and creating profiles of the murderers, which included describing their habits and attempting to predict their next moves.

Douglas profiled the "22-Caliber Killer" as a mission-oriented asocial loner with an assassin personality who had, in the past, joined hate groups or even groups with positive goals or values, such as a church, and was now convinced that he was contributing to the cause. His profile also suggested that the murderer had a gun fetish and military background and would have been discharged out of psychological issues or failure to adjust to military life; he was a racist who killed in a blitz-assassination style. Douglas did not think that the "22-Caliber Killer" was responsible for the cab driver deaths. Later, Christopher was indeed diagnosed as a paranoid schizophrenic. He claimed he was "ordered to kill" as part of a "conspiracy."

The whole episode brought back Nick's memories from a few years before when the Son of Sam terrorized New York City and its suburbs from July 1976 to July 1977. Everyone Nick's age was keenly aware of the Son of Sam murders as the media laid out his killings of young couples, many of whom were sitting in cars. The New York City area was on heightened alert with no one knowing exactly where he would strike next. David Richard Berkowitz, known as the Son of Sam, was also called the .44 Caliber Killer during the time that Nick was in college and graduate school. The Son of Sam murdered six people in New York City from 1976 to 1977, claiming he received orders from a demon-possessed dog.

Christopher was indicted by an Erie County, New York, grand jury in the first three .22-caliber killings, and on April 27, 1982, Supreme Court Justice Frederick M. Marshall found Christopher guilty of second-degree murder in all three slayings. The mystery of the cab driver killings has never been adequately closed.

Later in Nick's life, at the end of his career, another horrendous racially motivated mass murder occurred in Buffalo when an evil young man drove hours from the New York/Pennsylvania border to murder black people on the east side of Buffalo at a supermarket. Nick was reminded yet again that sickness and evil exists and that vile humans murder in the name of racism or some other satanic reason and society never seems to learn any lessons from the past as mentally ill or pure evil people walk our streets.

Chapter 10

GO WEST, YOUNG MAN, AND AS THE SALMON SWIM

I left the Roost after my usual interesting conversation with Marty D with a pizza and the glow of a bourbon and headed to the Gypsy Parlor right down the road. I walked into the bar and was greeted by Gabby, the owner, and Jon behind the bar. I ordered a Peroni while I waited for two chicken pastelitos to take home. When I left the bar with my food, I waited for the light at the intersection at the corner of Grant Street and Potomac next to the bar. As I waited for the light to change, an attractive young black woman wearing cut off jean shorts and a tank top walked across in front of my windshield. As she was passing, I noticed her talking intently on her cell phone when all of a sudden she tilted her head back fiercely and let out a loud laugh. Just before reaching the other side of the street, she tilted her head back again in a huge laugh.

As the light changed, I turned left onto Grant Street and headed for the Scajaquada Expressway designed by the celebrated urbanist Frederick Law Olmsted. Upon entering the expressway looking up Scajaquada Creek, I glimpsed the familiar towers of the Church of the Assumption. This road was originally the Humboldt Parkway when the original Olmsted design fit in perfectly with its surroundings before it was bastardized in the 1950s change that turned it into what Buffalonians refer to as the 198. When I first moved to Buffalo, I had to get used to how the locals referred to certain streets as "The 198" or "The 33."

After my dinner, I sat down to review my notes of the day so I would be fresh to work on my assignment in the morning. Earlier in the day, I found a passage of something Nick had written—parts of the few things that survived the generations in scrapbooks. It was just a brief couple of sentences about a trip he made with his parents in the early 1960s to see a circus. Only a few sentences, the material provided a glance into what NYC was like to a young boy at the end of the 1950s and beginning of the 1960s.

Nick wrote: "When I was a young kid, my parents took me to New York City to see the Barnum and Bailey circus in Madison Square Gardens. Pieces of memories of the walk through the city was seared into my mind as foggy disparate views as seen by a young boy of five or six all wide eyed and mysterious. Walking across many streets engulfed with people walking here and there, men in suits and all with hats I would later see in film noir movies. I never was around so many people or this much cement and asphalt all with a film of grit and dirt yet clean just the same. The humanity was moving in a sort of synchronous chaos, and my mom was holding my hand tightly so she would not lose me in the crowd.

I recall my dad telling me to avoid walking on the metal grates. I noted steam rising around some manholes as I walked past a drunk—'Bums,' my parents mumbled, as I glanced back at a man and another nested to him both lying in a store entryway along the sidewalk. Steam rose from the bowels of the city around them as they in their drunken stupor both absorbed the heat from below in the late fall night. I have never forgotten these vivid scenes of my first memory of New York City and recall seeing similar ones later in life walking by those poor souls lying in drug-induced semiconsciousness in the crime-infested urban areas of the 2020s.

Nick took the long flight from the Buffalo Niagara International Airport to Seattle with a short stopover in Chicago. It was his second trip to the northwest but his first sight of Mount Rainier. Looming above a sea of clouds, the mountain caught him by surprise as he looked up from his book out his window seat at the awe-inspiring view seemingly appearing out of nowhere and forever seared into his brain. It was summer 1983, and his mission was to meet at the Seattle E&E branch office and then head with a crew to the Western Processing site in Kent, Washington, in the Green River Valley area for an initial site assessment. Western Processing was a chemical waste processing and recycling facility that operated from 1961 to 1983 on a 13-acre site 20 miles south of Seattle in the commercial area of Kent.

Nick spent two days at the site, and each day his first sight looking southeast was that of Mount Rainier rising massively into the sky and past the clouds. The view never got old, nor did the look and feel of the pacific northwest: all green and lush but a little too wet. Some of the Pacific Northwest's largest industries sent a wide variety of chemicals and waste materials to Western Processing. Based on this assessment, the Environmental Protection Agency (EPA) ordered the company to stop operations and placed Western Processing on the National Priorities List (NPL), as one of the most contaminated sites in the Superfund program.

Nick's research prior to his visit included the initial EPA write up on the property, which stated the following: "The 13-acre Western Processing Company originally reprocessed animal byproducts and brewer's yeast. The business expanded in the 1960s to include recycling, reclaiming, treating and disposing of industrial wastes from over 300 businesses over its working lifetime. These included electroplating wastes, waste acids (pickle liquor and battery acid), zinc dross and flue dust from steel mills, transformers, waste oils, pesticides and spent solvents."

When Nick first arrived at the property, his eyes were drawn to a conglomeration of large aboveground storage tanks of different sizes and shapes that dominated the view. What made this odder was that these tanks were freshly painted in different colors, including robin's egg blue, pink, purple, black, and white. The "look" kind of reminded Nick eerily of the brownstones of the Painted Ladies he saw in the Alamo Square section of San Francisco months before on a different trip. Besides the bulk "painted lady" tanks, numerous drums and other containers were stored on the property along with areas of buried materials, open waste piles, and lagoons.

After the 1983 assessment by E&E, the site was permanently closed by federal court order because it was found that site operations contaminated soil, groundwater, and sediment with hazardous chemicals.

The initial investigation included analysis of over 160 soil and groundwater samples that confirmed that hazardous substances had been released into the environment, had contaminated the shallow aquifer, and had caused widespread contamination of soils at the site. The initial work by E&E and the regional EPA office began in late April 1983 and was completed in August 1983 when Nick's visit occurred. A year after Nick's visit, cleanup activities included the removal of approximately 4,700 tons of wastes from ponds, drums, and tanks on the site. Years later, more than 25,000 cubic yards of contaminated soils and sludges were removed from the site subsurface and a 40-foot-deep vertical barrier wall ("slurry wall") was installed around the site. The wall and a groundwater extraction treatment system was installed to prevent further contaminants from spreading from the site into nearby groundwater.

Every waste site Nick worked on or visited was situated in a fragile and complex environmental dynamic causing potential impacts to the area's most sensitive environmental media. The Western Processing site was no different. The site was bounded on the west by Mill Creek, which flowed in a northerly direction into the Black River, a tributary of the Green River, which became the Duwamish River before ultimately emptying into Puget Sound at Seattle. Located outside the 100-year floodplain but over an alluvial shallow aquifer, the groundwater table was 5 to 10 feet below ground surface (bgs). Contaminants migrated to Mill Creek prior to the installation of the slurry wall around the site. These included metals from sludges and liquid wastes; spent solvents; caustics, flue ash, and ferrous sulfide; pickle liquor; cyanides; zinc chloride and lead chromate; and waste oil from reclamation processes.

The surface cleanup of wastes was completed in November 1984, with the exception of a dioxin-contaminated oily liquid discovered in one storage tank. This was finally, in 1986, destroyed by successfully treating the approximately 6,000 gallons of dioxin-contaminated liquid with a potassium hydroxide, polyethylene glycol mobile chemical de-chlorination process. Investigations confirmed that over 95% of the contamination was located in the uppermost fifteen feet of soil. This was a highly contaminated site, and concentrations in soils were detected at up to approximately 141,000 ppm of lead; 10,000 ppm of PCBs; 53,000 ppm of total polycyclic aromatic hydrocarbons (PAHs); and 580 ppm of volatile organic compounds (VOCs).

With those levels in the soil, it was no surprise that extremely high concentrations of contaminants were also found concentrated in the shallow groundwater table to approximately 30 feet below ground surface. Groundwater concentrations of up to 510,000 parts per billion (ppb) of zinc, up to 5,400,000 ppb of total semivolatile organic compounds, and up to 1,346,000 ppb of total volatile organic compounds (VOCs) were confirmed. In July 1986, a plume of trans-1,2-dichloroethene was found to have migrated under Mill Creek and was detected in wells west of the creek. This is not uncommon for chlorinated solvent plumes, which are heavier than water and thus sink to the bottom of the water-bearing zone and as such can migrate below otherwise natural barriers that would incept other contaminants such as petroleum contamination that is lighter than and floats on the top of water.

Like most of Nick's trips, he spent a short amount of time working at a site only to leave for the next site; he never knew the outcome of the samples he took or how they were used to eventually clean up the site. It was always on to the next mess, the new circumstances, and a new set of chemical hazards. Before he was at the Western Processing site, he had been on some other nasty place in some other place in America. One of those other sites was the Toftdahl-Davis Farm site in Washington State.

Nick placed his hand on an old deteriorating wooden post that formerly formed a fence long ago probably to keep the cattle from the rural road that ran along what was now his work zone. In what was now a rural neighborhood with small well-kept homes spiking randomly off the winding road, his work site was a vacant grass-covered field. He was setting up a site perimeter using hazard tape and figured he would use this old post to wrap the tape around before placing other stakes farther along the perimeter. Just as he was about to yell instructions to the other team members, he felt this intense pain on the back of his hand.

As he looked away, about to bark the directions to the crew, he began wiggling the post to test its strength. Nick stopped in mid-sentence, his mouth frozen in the shape of the word he was about to speak. The pain turned him silent as its intensity shot through his hand instantaneously activating the pain centers in his brain. This was not a mere annoying pain; no, this was a debilitating pain stronger than any he had felt during his many injuries as an athlete. His head whipped around, a grotesque look now formed on his face as his focus was immediately drawn to the back of his hand, the source of the now pulsating pain.

The back of his hand had a dozen or more fire ants, all with their reddish-brown heads digging simultaneously into his flesh. Before his brain even acknowledged the scene, he had smacked the back of his hand against his thigh, followed quickly by using his other hand to wipe away any remnants of the ants from his skin. He backed away from the post onto the black pavement of the street. "Holy cow!" he almost yelled as his colleagues, who looked at him with questioning eyes having just witnessed his unusually erratic behavior.

Almost a month before going to Western Processing site Nick had made a previous trip in July 1983 to the Northwest when he was sent to the Toftdahl-Davis Farm site. He and a crew were to excavate metal anomalies found about three feet below ground thought to be buried drums in a vacant parcel across from a rural residential neighborhood. He went with a couple guys from Buffalo and met some others from the E&E Seattle office.

The Toftdahl drum site was about 15-acres in Brush Prairie, Clark County, Washington. Up to 200 drums of unknown material were reportedly buried at the site in the late 1960s and early 1970s. Although some drums had previously been removed, it was unclear how many may still remain buried or how much of the contents may have spilled into the ground. The Seattle office had previously completed a geophysical survey on this portion of the site that was closest to the residential homes. The survey showed what the geophysicist thought were drum "signatures," and these were depicted on a site map. In this case, the team used a magnetometer survey at the property because it "sees" metal objects below the surface. This geotechnical technique is used in the environmental field for locating buried steel drums, tanks, pipes, and iron debris.

During his time onsite, six drums were excavated by Nick and his crew, placed in overpacked drums, and sampled. High concentrations of metals and organic contaminants were found in the material in the drums and adjacent soil. Three private wells near the site were found to contain low levels of some of these same contaminants.

A lasting memory of Nick associated with his pacific northwest trips was salmon. In the east, when you drive around you see steak restaurants and burger joints. During Nick's time working the various sites in Washington and Portland, Oregon, he was struck by the number and variation of restaurants featuring salmon; from salmon steaks to burgers to some other variation.

There are five species of salmon in the Pacific Northwest; the king (or Chinook) salmon, the coho salmon, the sockeye (or red) salmon, the chum salmon, and the pink salmon. Their habitat extends from the North Pacific Ocean to the Bering Sea. Nick's observation about the abundance and variation of salmon restaurants was just one of those silly little observations made when one is in an unfamiliar place that offers unique regional specialties. It was no different than his observation of and painful withdrawal from the magnificent Cajun food in Louisiana or the unending miles of cornfields in central Ohio.

Nick went on and on one day at the job site about the many variations of salmon restaurants during one of their breaks while they all sat around cooling off from a session in Level C in the hot sun. One of the contractors whom Nick befriended, as was his way, heard Nick's discussion and brought some wild salmon he caught to the job site the next day. He brought fresh salmon recently caught and lightly cooked and some smoked salmon to share with the crew. It was the best salmon Nick ever had, and Nick was never previously a real fan of salmon.

Prior to his trips to the Northwest, Nick had kicked off his whirlwind year of 1983 with a trip to the Southern Pacific Railroad site near Sacramento, California, for two weeks in early February. The job was to complete a surface and subsurface soil sampling program across a former rail yard using a conventional drill rig. At each grid location, the drill rig would complete a boring, and in each boring the crew would complete continuous split spoon sampling.

Nick's crew was using standard 2-foot long, 3-inch outside diameter split spoons to collect shallow and deep subsurface soil samples. A split spoon sampler is split cylindrical metal barrel that is threaded on each end. The leading end is held together with a beveled threaded collar that functions as a cutting shoe that is pounded into the soil at the chosen depth using a 250-pound hammer from the drill rig. Once driven down to the desired depth, the spoon is retrieved, and the barrel is split open to reveal the soil sample.

Back at that time, it was standard protocol to clean the spoon between each use by following a step wise decontamination process that included the following steps:
- 1) Wash with a mixture of alconox detergent and potable water.
- 2) Rinse with potable water.
- 3) Rinse with a solution of 10% nitric acid.
- 4) Rinse with distilled water.
- 5) Rinse with acetone.
- 6) Air dry

Nick often thought back to those early days and the amount of exposure he must have gotten from the acetone and other chemicals he used in an attempt to take unbiased samples of the hazardous waste and to accurately depict the concentrations in the environment without carrying contaminants from one location to the next. This was accomplished using a series of wash tubs and cattle brushes, spray bottles, and bottle brushes. Nick often wondered what kind of exposure he got from the acetone having to frequently reach into a tub to retrieve a spoon; the acetone made his hand feel cold through his nitrile gloves.

For more than 100 years, workers at the Central Pacific Railroad, and later the Southern Pacific, built and repaired engines and rail cars at the downtown Sacramento railyard. Those types of operations at rail yards resulted in the use of large amounts of chemical solvents, fuels, and lead paint, some of which sloshed to the ground. At this site, they reportedly disposed of the waste by piping it into a big lagoon and letting it settle into the soil. Nick spent most of his time taking and packaging soil samples for shipment to the laboratory and cleaning the spoons in between samples.

During most of the site visits, the crew had little time or energy for sightseeing or fun. Typically, they worked 10–12-hour days in some level of protection doing some manual labor. They were long hot/cold days, and by the time the day was done they just dragged themselves back to the hotel room. Usually, they tried to take the whole crew for a really nice meal one of the evenings.

On this particular field event, they all decided to go out to Old Sacramento, which was fairly close to the work site. Historically, Sacramento's waterfront has been described as "the site of both tragedy—fire and flood—and triumph—the raising of the streets, the founding of the Transcontinental Railroad, the terminus of the Pony Express, and the home of California's first thriving business district, fueled by gold, agriculture and the river."

At that time, Old Town Sacramento was a fairly new construction of a few blocks of town that was built to resemble an old west town and the history of the place. In the mid-1800s, this part of town was sort of the wild west full of gold rush miners, merchants, and madams. In 1848, California pioneers Samuel Brannan and John Augustus Sutter Jr. built a town where the American and Sacramento Rivers meet. That same year, gold was discovered at Sutter's Mill nearby. Prospectors soon arrived in droves on their way to the gold fields. The city now called Sacramento became California's state capitol in 1854.

Over the next century, the commercial district moved away from the river to avoid floods. Soon, the old part of town became known as the worst skid row west of Chicago. Redevelopment began in the 1960s, and 28 acres of land in Old Sacramento became the first historic district in the western United States.

The railroad extends from the museum property located in Old Sacramento south along the east bank of the Sacramento River levee. The original Sacramento Southern Railroad ran south 24.3 miles to Walnut Grove, California, via Freeport and was a nonoperating subsidiary of the Southern Pacific Company incorporated in 1903. The line was constructed between 1906–1912, and the first train began operating over the line in 1909. Nick's lasting memory of that dinner was the dessert. He had the best apple pie covered in French vanilla ice cream he could remember having, certainly rivaling any of his grandmother's which until then had surpassed all others.

Chapter 11

A Senseless Death

Nick walked into his apartment in a complex of two-story brick apartments located just off Transit Road and was surprised to see his roommate Mark. It was Friday the 13th in the month of June in 1980. The apartment was behind the bowling alley along Transit Road just north of Sheridan drive in the Village of Williamsville, New York. Transit Road ran from Lockport to the north towards Orchard Park and East Aurora to the south. It is the dividing road between Clarence, New York, to the east and Williamsville and Amherst, New York, to the west.

"Hey, Mark, welcome home," Nick said as he walked past Mark, who was prone on the couch as he proceeded on his way to his room to change. Nick lived with two fellow colleagues at E&E, and the three of them rented this apartment not far from their office near the airport on Sugg Road. The three were going to spend the weekend moving to a house in Clarence on Thompson Road that they were going to rent from another colleague from E&E who had recently been transferred to Oklahoma for a long-term assignment.

Mark replied, "Yeah, glad to be back from that project I'm working on in Montana for my two-week rotation home before I go back out again." Mark was an aquatic/fisheries biologist or generally a terrestrial/aquatic field biologist. His expertise was in performing surveys for birds, mammals, and herptiles. Also known as herps, herptiles are reptiles and amphibians. Their other roommate Mike was a PhD archeologist. Nick was an environmental/hazardous waste scientist. A third colleague, Kit, was a geologist and would eventually move into the house they rented all together in Clarence.

It was not uncommon for one or more of them to be gone for weeks or longer at a time on some field project somewhere across the country. Mike also had a number of projects in Puerto Rico and on Vieques Island, better known as Isla de Vieques, off Puerto Rico in the northeastern Caribbean. Vieques had a 400-

year history of Spanish and before that native influence. The Navy used this island as a bombing range and testing ground.

Early in Nick's time at E&E, Mark and he became friends and decided to share the apartment not far from work in Williamsville a block back from Transit Road on the border of Williamsville and Clarence, New York. Not long after, Mike moved in because he needed a place to stay. The three, although hugely different in personality, became good friends. One time at work, out of nowhere, Nick decided to call Mark up and pretend to be a professor from some big university wanting some input on something from Mark. Well Nick told Mark that he was Professor Erwin Cory (a comic on TV at the time) and went on and on for a few minutes with Mark acting serious and not catching on. Nick, not being able to maintain the gag, started to break out in laughter and all Mark said was "I'm going to get you," and that was that.

About a month later, Mark got his revenge. Nick had bought some coloring books for his nieces and nephew for when they came to visit, and some of them had been colored outside the lines and haphazardly as small kids do. Nick was making an important trip to the Washington DC office to go over all the equipment purchases and other things about the special project fund he was administering. He had this large accordion brief case he got from work to bring all the brochures and info for the meeting.

At the large opulent Washington office conference table with all the E&E Washington office higher ups that ran the TAT and FIT EPA contracts, Nick reached in his brief case and pulled out all the material he wanted to pass out and discuss. He threw the material down on the table, not looking at what he was grabbing, and with a swoosh out came all the brochures across the table with the Mickey Mouse and Donald Duck coloring books on top with their pages opened to the childishly colored pages. The head of the program—Rodger, who Nick was trying to impress—just looked at Nick and said, "Did you have a fun airplane trip coloring?" Yeah, Mark had gotten Nick back in spades, and they had some good laughs over that over some beers.

Mark's assignment in Montana was associated with clearing the land for an eventual oil pipeline, and on this particular tour he was studying a large colony or "town" of prairie dogs located in the pipeline route. Prairie dogs are named for their habitat and warning call, which sounds similar to a dog's bark. After his assignment in Montana, which was winding down, Mark was eventually assigned to a different section of the northern border project—a pipeline project. Because he was also proficient in aquatic biology, he was needed to assess the various pipeline stream crossings along the pipeline route. Much of the of the work was in North Dakota. North Dakota was on the eastern U.S. leg of the Alaskan Natural Gas Transportation System.

The Northern Border Pipeline was scheduled to be in operation by the fall of 1982. When completed, the pipeline would carry surplus Canadian natural gas and eventually Alaskan natural gas in the late 1980s to the Midwest and eastern United States. The initial phase of the pipeline stretched 823 miles from Port of Morgan, Montana, to Ventura, Iowa, and ran through parts of Montana, South Dakota, Minnesota, and Iowa. Later in Mark's career because of his unique and extensive experience, he become a fulltime pipeline worker on the various lines that crisscrossed the United States.

Mark yelled to Nick in his room, "Bruce is stopping over after he leaves the office, and we are going to drink some beers and just hang out." Bruce Garlapow was a senior wildlife biologist at E&E and somewhat of a mentor friend to Mark. Mark first met Bruce as an undergraduate at SUNY Fredonia while Bruce was there completing his Master of Science thesis on the reintroduction of wild turkeys in Western New York. Dr. Kevin Fox was both his and Mark's advisor at SUNY Fredonia. Mark got the job at E&E because of Bruce when a position came open for a field biologist to assist doing a wetland study for the Utica North Thruway Interchange. Mark continued, "We wanted to catch up while we are both back in town, and Bruce mentioned that he had to catch a flight to Houston for meetings on Monday."

...

"Jedidiah," yelled my friend Chantel as I was leaving her house after a short visit. "You going to be back tonight to meet up at the Gypsy Parlor or McCarthy's?" I kind of yelled something back to her as I headed for my car about getting back to her later, and off, I went. Late that afternoon, I came up over a rise in the road on Route 83 just past its intersection with Creek Road in Arkwright when I saw the windmills come into partial focus through the mist, which seemed to almost envelop the windmill off to my left. I had just completed a wonderful kayak trip in the Chadakoin River just outside Jamestown, New York, and decided to drive back through Arkwright on my way back to Buffalo so I could get a feel for the history of the murder of one of Nick's colleagues at E&E. This was all part of research to fill in the blanks of Bailey McCarthy's ancestor Nick McCarthy. Delving into the history was my passion, and I was eager to get some background on Arkwright, New York.

"The wind turns the sails, and the sails turn the millstones" was the thought I had when first seeing these monsters come into view as I turned the corner in the road. But these were not the windmills of yesteryear or those that invoked thoughts of Don Quixote. No, these are much more formidable foes than the ferocious giants Don Quixote battled to collect the spoils and the glory as a knight. These multi-megawatt-producing metal monsters—with their huge metal towers with three offset blades—look so alien in juxtaposition with the memory of the wooden Dutch masterpieces that always seemed to be one with the land and the green farmland that surrounded them. I almost got the feeling that I would be propelled into some spaceship with small ghostly white beings with large lifeless black eyes. Ok, I thought, these are certainly more appealing than the massive number of large powerlines I saw in Niagara Falls—well, maybe.

The forty-seven wind turbines are located within the Town of Arkwright, and the project connects to the grid in the Town of Pomfret. I had just read that some citizens were suing the developers of the wind farm. They were asking for unspecified damages related to "loss of property values, compensatory damages for destruction of homes and lifestyle, loss of use and enjoyment of their properties, damages for relocation costs and time spent relocating, mental anguish, destruction of scenic countryside, physical pain and suffering, difficulty sleeping, nuisance, trespass, interference with electronics in their homes such as satellites, telephones and televisions, loss of business profits, special damages for stress, anxiety, worry and inconvenience, and the effects lights and noise from the turbines have on their properties."

The lawsuit further claimed "Plaintiffs allege that they own and, therefore, have a lawful right to possess the real property on which they live or own. Plaintiffs allege that the giant wind turbines that defendants have placed around their property results in a trespass by the defendants due to invasion of their land by noises, lights, flickering, and low-frequency vibrations which penetrate their homes, thereby destroying the use and enjoyment of the plaintiffs' land; among other trespass."

Arkwright is hilly land located in the northeast part of Chautauqua County, New York, and is just southeast of the adjacent but different municipalities of Fredonia/Dunkirk, New York, the latter of which is located on the magnificent shores of Lake Erie. Thirty-six square miles, Arkwright was settled around 1807 and established in 1829 from the towns of Villenova and Pomfret. It is just a beautiful place, highlighted by Canadaway Creek with its deep gorge and 22-foot waterfall that runs across from East to West before spilling into Lake Erie.

The stream was originally settled by the Erie tribes and later by the Iroquois, who called the stream "Ga-na-da-wa-o", meaning "running through hemlocks." The early European settlers from Eastern and Central Pennsylvania ended up pronouncing the name as "Canadaway." The Native American name probably referred to the dense canopy that still covers the deep gorge at its headwaters. Early surveyors named the creek "Cascade" after the scenic falls that are located in the town of Arkwright. The first nonnative settlement along its banks occurred in 1804 and was called Canadaway, which later became the Village of Fredonia.

The mouth of Canadaway hosted the first naval battle in the War of 1812 when an American military company held off a British gunboat as it tried to seize a salt boat from Buffalo that had sought sanctuary in the creek. Today, the area contains large nature preservation areas; a 33-acre Canadaway Creek Preserve and the Canadaway Creek Wildlife Management Area. Both were designated to protect around 140 species of birds, including the yellow-bellied sapsucker and great blue heron and also to protect the headwaters for its importance to the local fisheries, which have distinctive physical characteristics and spawning habits.

As I devoured the historical background, one particular bit of history caught my attention; the history of the Chicken Tavern. Not far from Black Corners and Black Pond was the location of the famous Chicken Tavern. It was first built as a log cabin on the summit at the intersection of the old Indian trail that ran south to the Allegany and the trail connecting the Canadaway settlement (Fredonia) with the Chautauqua Road over in Little Valley. There is really nothing there today, but Chicken Tavern was where today Route 83, a road that links Fredonia with South Dayton, crosses Creek Road in the Town of Arkwright.

The town is named after Richard Arkwright, the inventor of a spinning device. But for many years, after the lumber trade, farming was the predominant craft. By the 1860s, butter- and cheese-making was dominant, and long, one-story "cheese" buildings spotted the countryside. Kept in these "long houses," the cheese was rubbed and turned daily to cure it properly. At that time, Susan Skinner became the champion cheese turner, making one dollar a day turning and rubbing the precious product in these cheese houses until the weather became cold. Cheese buyers from New York City would come and often buy the entire lot at a price that varied from 5¢ to 10¢ per pound. Arkwright was one of the first towns in New York to form a cheese cooperative.

The town cemetery and the famous old Chicken Tavern of Arkwright have a common bond. They once belonged to the Town family. In January 1826 Asa Town, born 1770, son of Asa and Eunice Town, migrated to Chautauqua County with his wife, Sally, and their six children: Aaron, Francis, Amos, Joel, Lyman, and Betsey. They bought a portion of lot 12 from Bethuel Harvey, which Bethuel had bought from the Holland Land Company. Located on the western edge of lot 13 in Arkwright, it's somewhat a mystery when the Chicken Tavern was built.

Some say it was built in 1818 and some say 1822. Others say it was actually built in 1826 by Aaron Town, who owned a log house on the site and operated a tavern there for a few years prior to 1826. In 1826, Aaron reportedly tore down the house and erected the structure later called Chicken Tavern. The Chicken Tavern, originally a big old log house, was replaced with a large, spacious post-and-beam structure built somewhat along the lines of a New England inn.

Both a tavern and an inn, it was widely known for its food and for its dance floor. It was the scene of many parties, dinners, dances, and town meetings. There was even a blacksmith shop on the property. The tavern was a stopover place for the stage route from Lake Erie to points south where horses were exchanged for fresh ones only to be changed for again on the return trip. Accounts suggest that "Passengers paid a fare of six and a half cents a mile and were allowed fourteen pounds of baggage free. In summer, the carriages would roll along easily enough, but woe betide the unwary traveler at other seasons. The male passengers were required to aid in raising the wheels out of each mud hole into which they sank, often to the axle."

Chicken Tavern was noted for its magnificent cooks. In its early history, it reportedly had a Dutch oven in the yard to do the vast amount of cooking for the family and the tourist trade. The inn was of course famous for its chicken dinners, and the dance floor was in use well into the 20th century. Asa's son Aaron Town ran the tavern for many years and handed it down to his sons and their sons. Oliver was the last owner. He died in 1901, and his widow lived at the tavern until she died in 1931. The tavern then became deserted, rapidly deteriorated, and collapsed.

Of particular interest to me was that the Town Tavern was also reportedly part of the underground railroad back in 1858, when slaves were escaping to Canada. A man named John Little took charge of escaping slaves through the tavern. John Little, his wife, and several children are buried on their farm in a little cemetery long since lost. But his sons William Patrick Little and Riley L. Little served in the Civil War and are buried in the Town Cemetery behind where Chicken Tavern once stood. William was the last buried in the cemetery.

Before I focused back on the tragedy that happened in Arkwright, I stopped for an ice cream cone staring out blankly, my mind recalling my kayak trip earlier in the day. Serendipitously, on the day I decided to further the process of my deep study of Nick McCarthy and a tragic event from his early career, I found myself instead off on a long kayak trip probably wanting to avoid the story I was about to research. It did not hurt that much later in Nick's career he worked on a site in Jamestown. Was it my knack for discovery or just a combination of accident and a need for a diversion that led me to Jamestown on that day? Pursuing some mental cleansing, I discovered what would become one of my favorite places to kayak. My assignment was to focus on Nick McCarthy's career beginnings; however, at a spot that I left off on during my earlier project, I had found that later in Nick's career he had worked on a Brownfields site on the Chadakoin River off Water Street in Jamestown, New York. This led me to find this memorable kayak location while looking for a sense of Nick.

Started by the federal government in the mid-1990s, comprehensive Brownfields legislation was signed into New York law (the "Act") in the 2000s. The Act created a Brownfield Cleanup Program ("BCP") with the goal of encouraging private-sector cleanups of brownfields and to promote their redevelopment as a means to revitalize economically blighted communities. The project Nick worked on was situated on the Chadakoin River about 1.2 miles southeast of the southeast end of Chautauqua Lake. The City of Jamestown is located in southern Chautauqua County, New York. Situated between Lake Erie to the northwest and the Allegheny National Forest to the south, Jamestown at that time was the largest population center in the county. In the late 1800s and early 1900s, a number of worsted textile mills were located in the city. Worsted is a high-quality type of wool yarn, the fabric made from this yarn, and a yarn weight category. The name derives from Worstead, a village in the English county of Norfolk. Jamestown was also known for furniture manufacturing and was once called the "Furniture Capital of the World."

It started early with my thought of going on a trip I had planned a year before but forgot about until today. The drive down to Jamestown, New York, from Buffalo on a late August early Sunday morning brought me to the Water Street project site after about an hour-and-forty-minute drive. The water was too low to kayak at what was Nick's project site location, so I proceeded to Lucille Ball Memorial Park in Celoron, New York, at the southern end of Chautauqua Lake.

I found the kayak launch along the lake but quickly realized this was not the day to be on the Lake in a kayak as the wind speed was at least 15 miles per hour with white caps showing. I made my way south a few miles up the Chadakoin River and found a brand-new kayak launch at McCrea Point Park.

The Chadakoin River is a 7.8-mile-long stream that begins at the southern end of Chautauqua Lake. Because of a geologic formation, the Chautauqua Ridge, the water flows eventually to the Gulf of Mexico rather than the nearby Lake Erie following the Chadakoin River to Conewango Creek and eventually to the Allegheny River.

As the City of Jamestown became more populated, the area along the Chadakoin River as it meandered through the city became industrial. The river below Warner Dam was hidden by factories and sections of the river were covered by buildings and the river became polluted in the urban area. The first few miles of the river, from Chautauqua Lake before Jamestown, are referred to as "The Outlet" and have been continuously used for water recreation. The large marsh areas along this length of the Chadakoin are home to many birds and other animals.

This beautiful small river full of lily pads alternating along both river sides as the river twists and turns and meanders along was just a wonderful discovery. I paddled upstream from McCrea Park towards its start at Chautauqua Lake, and with each river turn there were more beautiful scenes to see. I was able to find one of the two steamboat hulls from the early 1900s and passed a few great blue and green herons, beaver, deer, kingfisher, kingbirds, and other wildlife along the way. I was spellbound by the natural, seemingly untouched beauty as swamp land, brush, and mature trees hugged the shoreline with no indication of human activity until I got closer to the lake.

Nature's artwork was on display as I passed distressed, twisted tree branches or the transformation of an intricate root system of a long-ago overturned tree into a natural wooden sculpture; the wind, water, and seasons created tangled and twisted shapes of nature better than any wood carver could conceive. Abstract shapes of driftwood with creature-like forms seemingly popping out here and there along the riverbanks.

As I took a rest from my steady pace, determined to stay hydrated and grab a drink of water, I looked down at the crystal-clear water. I noticed the river was not as deep as I thought just as I spotted a few yellow perch undisturbed by my stationary kayak. As I began gliding along again, letting the current take my kayak and not padding, I could see swarms of young of the year fish, which flashed away quicker than a blink of the eye when they saw my boat's shadow. Undoubtedly, they believed some ominous predator was coming.

I made the source of the river at Chautauqua Lake just as the steamboat the Chautauqua Belle passed me as it was returning from its two-hour historical tour back to its mooring at McCrea Park. The steamer Chautauqua Belle is an authentic replica of a Mississippi River–style sternwheel steamboat that was used to shuttle tourists across the lake for a historic tour.

I turned around just as I entered the lake but not soon enough. The wind coming across the lake made for a very rough ride that almost capsized my kayak as I turned perpendicular to the waves. The trip back allowed me to focus on sights I missed on the trip up, and it just made the whole day one of my more memorable kayak trips. About 45 minutes later, I had finished tying my boat down in my truck and began to refocus on my side trip to Arkwright.

Bruce and Nancy Garlapow lived in Arkwright, New York, not far from the former location of Chicken Tavern on a property adjacent to the State Forest lands. Bruce had helped Mark get the job at E&E. After Mark worked on that initial wetlands project, he was hired full-time and went right into the Northern Border project. As noted, the Northern Border Pipeline was a major petroleum transportation system that linked the Midwestern U.S. with reserves in the Western Canadian Sedimentary Basin.

The initial phase of the pipeline stretched 823 miles from Port of Morgan, Montana, to Ventura, Iowa. In addition to transporting Canada-sourced supply, the Northern Border Pipeline received and transported natural gas produced in the Williston and Powder River Basins in the United States and synthetic natural gas produced at the Dakota Gasification plant in North Dakota.

Mark walked into Nick's office sometime in the middle of a rainy Tuesday morning a couple days after the weekend with a worried look on his face. Nick was about to ask his normally jovial roommate what was wrong when Mark blurt out, "Bruce never showed up for his meeting in Houston. Nobody's heard from him, and no one is home at their house. Worse, Nancy didn't show up to her job as a social worker at the J. N. Adam Developmental Center in Perrysburg either. I'm going down to check on their dogs; my aunt lives about five miles from Bruce and Nancy, and I'll try to find out what's going on."

Nine days later, the senseless reality of the story eventually unfolded, and to the shocked employees at E&E it was surreal to say the least. It appears that the next day after hanging out with Mark, Bruce and Nancy were picnicking in the Canadaway Creek Wildlife Management area when a local teenage kid with a .22 rifle shot him in the head "thinking he was some varmint moving around in the bushes," as Mark described it. Bruce was dead immediately, but Nancy took multiple shots after the kid panicked.

Kenneth Wilbur of Bard Road, Town of Arkwright, was 19 when he committed the murders. Wilbur contended he didn't mean to shoot Bruce when his .22-caliber rifle he was using to hunt woodchucks discharged accidentally as he startled the couple. He then reportedly shot Nancy twice in the head to silence her screams. At some point he also killed their dog before he hid the bodies under nearby brush. Nine days later, he confessed the slayings to police and led them to the bodies.

He was tried on two counts of second-degree murder; however, a Chautauqua County Court jury instead found him guilty of first-degree manslaughter and criminally negligent homicide. County Judge Lee Towne Adams sentenced Wilber to nine to 27 years in state prison. Disappointed by the lesser sentences, the Chautauqua County District Attorney declared at the time: "Wilber's testimony was at variance with physical evidence in the case," describing Nancy's killing as "an execution-type slaying." He said psychiatric reports on Wilber indicated he is "a dangerous individual." In his view, "Wilber should serve the maximum sentence."

The family mounted a petition campaign each time Wilbur came up for parole, and the state Division of Parole rejected parole for the convicted killer three times. The District Attorney had told parole officers, "Kenneth Wllbur took the lives of two young people who were enjoying a picnic in the wilderness they dearly loved. They were cut down in the prime of life by an extremely dangerous individual. For the sake of safety of others, I most urgently request that parole be denied." Mystery still swills around the murder, and many familiar with the facts, photographs of the murder site, bodies, bullets, and autopsy reports think the true story has never been told.

To Nick and Mark and his other colleagues at E&E, Bruce was an honest, highly intelligent, and hard-working friend. Various benefactors established "The Bruce and Nancy Garlapow Memorial Scholarship" at Fredonia College in honor of Bruce, a 1974 Fredonia biology graduate, and his wife Nancy Kulick Garlapow, a 1975 Fredonia sociology graduate. The award was and may still be given to incoming freshmen students expressing career interests in biology or environmental science.

Chapter 12

Petro Processing – Baton Rouge

I put in at one of my favorite places to kayak at a small launch in a little known park in Tonawanda, New York, called Eastern Park at 280 Fillmore Ave. I liked this hidden away place because it was little used, small, and very quiet and even better has an EZ-Lock dock with rollers especially designed for older kayakers like me because of how easy it makes getting into and out of your kayak without the fear of rolling over into the water. The only drawback was for some reason the town had let the launch go, and it was being held together in place with zip ties and wire.

What I especially liked about this launch was its location on Ellicott Creek adjacent to two old historic rusted steel train trestles that provided a supreme backdrop to a series of boat houses hugging the shore to the east as the creek flowed from Amherst and Clarence through parkland on its way towards the Erie Canal, where Tonawanda and North Tonawanda shared the canal just before it emptied into the Niagara River. Going west from the launch meant paddling under the bridges and around their metal pillars long ago sunk deep into the creek floor. Past the massive steel structures are some small commercial buildings intermingled with houses along the shore before paddling past boat docks and small creek side restaurants. These just up from its mouth with the Erie Canal and the Cities of Tonawanda and North Tonawanda, which nestled around this confluence.

What made this location and paddle even more relevant to my research was that one of the restaurants was run by Nick's wife a year or two before her battle with the disease that took her life. The dock and patio along the creek enticed boaters and kayakers in for lunch or an early dinner. Looking up from the water at the restaurant as I let the current drift me onward, I could almost imagine her two standard poodles that she brought to the restaurant daily outside frolicking along the shoreline.

∙ ∙ ∙

"Nick, you feeling ok?," one of his crewmates asked as Nick seemed to stumble with his first step. They were on their way back to the well where the weird reaction had occurred. They had abandoned their efforts the day before until they understood what they had seen in the sample bailer the day before. After consultation with the chemists back in Buffalo and some of the other corporate brain power, they had devised a course of action to safely obtain a sample from this well that contained a weird mixture and had precipitated a reaction none of them had ever seen.

They were about a week into their assignment. Although Nick had made a special effort to alternate the crew's downrange time, he had forgot about himself. He thought he had to be the one going downrange for safety reasons. His poor choice was now coming back at him in a drastic and fast way. Even though he had been pouring Gatorade down his throat daily and demanding others do so also to replace the electrolytes they quickly lost each day, he was going down fast. He knew he was in trouble and immediately recognized the symptoms of heat stress as his body was quickly failing him. He motioned to the crew to get him out fast. Nick was helped to the back of the truck and driven immediately to the decontamination line. By the time he reached the first decon tub, he was already over-breathing his respirator because he could not get enough oxygen. His pulse was racing, and he felt a cold sweat in the hot sun. Nick barely recalled getting through decontamination because he almost passed out and had to use all his energy to stay conscious. He was a stubborn guy.

Five minutes later, he lay on the stretcher outside the field shed under a tarp telling the others he did not have to go to the hospital. "Well, buddy," Andy said as he looked down at him, "unless your pulse decreases soon, you are not going to have much of a choice because I'm not about to see you leave here in a body bag." Luckily, Nick regained his normal pulse fairly rapidly once his electrolytes were restored while lying in the shade. Looking back, the crew all decided that he should have gone to the hospital. It was three days before Nick went back downrange. It was a mistake he would never make again, and he often retold the story at safety briefings for years afterward.

∙ ∙ ∙

It was Saturday, October 1, 1983, and Nick had just gotten off the phone with his parents, whom he had called so he could wish his mom a happy birthday. He only saw his parents during Thanksgiving when he made the trip back home to Peekskill and then again once during the summer, usually at their vacation home in Cape Cod.

He spent the next few hours packing for his trip to Baton Rouge, Louisiana, to go to a site named Petro Processing. His team from Buffalo would meet up with a few others from the E&E Texas Field Investigation Team (FIT) office and some guys from their partner company CH2MHill. The E&E Buffalo team included Nick, who was to be the lead Health & Safety Officer and senior technical guy along with Jim, Paul, Glenn, and Andy. Even James B. Moore "Himself" was coming later to help in the deep well sampling. He'd be bringing some special equipment and a tripod to get the work done.

Their mission was to find and sample a series of groundwater wells that had been installed a number of years before. They had to get this work done, collecting sufficient quality data, in a two-week period to support the EPA, which was in litigation with responsible partners or potentially responsible parties (PRPs). PRPs are defined as "any individual or organization—including owners, operators, transporters or generators—potentially responsible for, or contributing to, a spill or other contamination at a Superfund site." In this case, it was a laundry list of some of the major oil and petrochemical companies in the United States.

The team was also responsible for collecting samples from residential drinking water wells nearby. What made this assignment even more pressure filled was that the "responsible parties" had hired a third-party individual to watch every step the crew took looking for opportunities to throw question into the data the crew would collect. Nick thought, "It's hard enough collecting samples from old wells lost in the wilds of the Baton Rouge bayou in Level B personnel protection without some third party looking over your shoulder watching your technique every minute of every day." This particular individual was slick, making friends with crew members and offering to take them out to lunch.

Nick had to remind them all that this was just a trick. Had they accepted the invitation, it would be used against them in the upcoming legal proceedings, calling into question the integrity of the crew. The PRP-hired JAFO (just another fucking observer) flashed a smirk when Nick informed him that it was not acceptable for him to be buying the crew lunch and that although he seemed like a nice guy, this tactic was not appreciated.

The Petro Processors of Louisiana Inc. Superfund Site, as it was later designated, was located near Scotlandville, East Baton Rouge Parish, Louisiana, about ten miles north of the City of Baton Rouge. The site contained two petrochemical disposal areas on the banks of Bayou Baton Rouge, which was in the floodplain of the Mississippi River. Bayou Baton Rouge meanders around the site and eventually into Devil's Swamp. The Bayou historically ran through the site but was rerouted during early remedial activities. The site was used for petrochemical waste disposal from 1961 to 1980. In 1980, the United States EPA, the State of Louisiana, the City of Baton Rouge, and the Parish of 1 East Baton Rouge filed suit against Petro Processors of Louisiana Inc. as well as several other generators who disposed wastes at the site.

Early indication was that toxic organic compounds and heavy metals had been released into local waterways and were posing a threat to an underground drinking water supply. A final site assessment decision was made on September 9, 1983. A Consent Decree (CD) for site closure was developed with the participation of all PRPs and it was entered into the Federal Court's record on February 16, 1984. Based on investigation results, the site's principal pollutants were found to be petrochemical wastes, including chlorinated hydrocarbons (hexachlorobutadiene and hexachlorobenzene), polycyclic aromatic hydrocarbons (PAHs), heavy metals, and oils. Groundwater, surface water, and sediments were all impacted.

The land near the facility was predominantly rural and sparsely populated. The population within a one-mile radius of the site was approximately 100. The nearest residence was about 3,000 feet from the site, and it was Nick's crew's job to sample some of the nearby drinking water wells on people's properties. Nick was told by some of the Texas crew that he should be very careful approaching the nearby residence because in these rural parts, people may not take kindly to strangers dressed in funny clothes and speaking in a northern accent.

At one particular residential property, Nick sauntered up a dirt driveway and around a bend to behold an interesting site. There were chickens and ducks and pigs and other farm animals just freely wandering around the yard, which contained a small well-kept farmhouse. The property was owned by an African American family who came out to greet this stranger with smiles and pleasantries. The only problem was that Nick could not understand one word being spoken because of the deep and heavy Cajun dialect and unfamiliar phraseology. Nick did not know how, but somehow he managed to convey to these people that he wanted to sample their drinking well, which was in a fenced-in area in the middle of the yard. He managed to get the samples and leave without understanding one word, and he never knew if they understood him either, but he recalled what friendly, outgoing people they were.

When they first arrived at the Petro Processing site, they unpacked their gear and equipment for the two-week job and set up the decontamination zone along the former access road to the site as close as possible to the contaminated zone. Photo31-33 This was done to reduce their time getting on and off site. The health and safety plan called for Level B personnel protection, which included chemically resistant suits and a couple layers of boots, booties, and gloves along with a self-contained breathing apparatus with supplied air that would last 20–25 minutes depending on the rate of breathing of each individual. It was hot—about 80 degrees and almost 90 percent humidity—and wearing this level of protection would be a unique safety hazard.

Protecting against heat stress was on the top of Nick's list of concerns, and so they erected a series of makeshift tarps for shade all arranged in a row leading to a rented shed that they would use as a command post and equipment storage. Their first task was to locate the monitoring wells that had been previously installed in and around the site. The problem was they only had a general idea where the wells were, and the site had become massively overgrown since they were installed.

The people of Louisiana have been nurturing and utilizing native bamboo for thousands of years in the low-lying lands around Baton Rouge. Arundinaria gigantea, or river cane, remains integral to the native American basket-weaving traditions in the region. River cane is a running bamboo that grows in thickets referred to as canebrakes. As Nick and the crew walked along an overgrown path, their advance was impeded by bamboo, which closed in all around them as they shuffled along in their suites. Their vison was greatly reduced by their full-face shields. You would have thought that one of them would have remembered the dark stories from Vietnam that blasted their TV screens nightly when they were young in the 1960s and early 1970s. It really was not that many years ago when nightly newsreels blasted stories of Americans killed or maimed by booby traps.

Later, movies were madw that seared into their brains the ugliness of the Vietnam war which came into their living rooms and onto TV sets. The Vietcong made traps with sharpened bamboo stakes, often smeared with urine, feces, or another substance that would cause infection in the victim.

As safety officer, Nick completed a safety briefing before they took off on their first walk. It covered all the safety concerns of the chemicals and their protective clothing—and of course about the heat stress, venomous snakes, and insects. He remembered to cover the safe use of and concerns about machetes, impressing upon them that no one should swing a machete unless the person nearest them was far enough away. Forty-eight species of snakes can be found in Louisiana, but only seven of them are venomous. This is where the Texas crew came in handy. They knew the snakes to avoid and brought snake guards with them, which they wore covering their lower legs. They rattled off for the crew the snakes of concern, including the eastern diamondback rattlesnake, timber rattlesnake—also known as the canebrake rattlesnake—the pygmy rattlesnake, the eastern copperhead, two different types of coral snakes, and the northern cottonmouth water moccasin. They also reminded the crew about the venomous spiders, including the brown and black widows and the brown recluse. Of course they also covered the American alligator, although most of the crew's time would be spent on higher ground.

As they walked blindly along the path, the front part of the crew cutting through the thicket of bamboo in search of the wells, they had little thought about the consequences of their actions. They were more worried about the venomous spiders and snakes—until the next day. The next day, they shuffled down the same path, moving farther across the site in search of the wells. What they saw along the path were dried sharply cut bamboo spikes sticking up just begging for one slip and fall to impart their revenge for being so unceremoniously cut the day before. Nick learned a lesson that day—a lesson he was sure that most local survey crews already knew—that he was sure to include on the many fields jobs he would be on in the future; the proper way to cut and leave foliage bamboo or otherwise prevent a trip resulting in puncture wounds was always covered in future safety briefings. When cutting, never leave a spiked or sharp edge that can cause a puncture wound or worse if a trip and fall occurred and leave the cut high enough when possible, to lesson a fall onto it.

After about three days, the crew had located all the well groupings on various parts of the property, and now it was time to start the sampling. Most of the well groupings were located in nearby thickets that the crew had cleared just off the access road that meandered around the site. On this particular day, almost a week into the job, Nick and the downrange crew jumped in the back of the pickup truck for the drive to the far end of the site, where the worst contaminated wells were. They brought bolt cutters to cut the locks and disposable bailers to collect the groundwater. A bailer is a portable grab sampler typically used for retrieving groundwater samples from monitoring wells. Bailer samples are collected when the bailer is attached to bailing cord and lowered slowly and gently down the well until the top of the bailer is below the groundwater surface. The bailer is pulled up when the desired depth is reached, with the weight of the water closing the check valve. Bailing line can be cord or cable and is hooked or tied to the top of the bailer for lowering it down the well casing.

As the weight of the bailer causes it to begin to sink into the fluid, the hydrostatic pressure of the fluid pushes up on the check valve (usually a ball check) at the bottom of the bailer, causing the valve to open and water to flow into the tube. Water entering a bailer fills until the level inside the bailer reaches the level outside the bailer. When the bailer has filled to its submerged level, the valve closes, preventing water from escaping. The bailer is retrieved using the tether, and the sample is discharged to an appropriate laboratory sample container.

The most common bailer material is high density polyethylene, known for its chemical resistance, strength, and low cost. Other materials include PVC, Teflon, and stainless steel. When Nick sampled wells at the beginning of the hazardous waste industry, he used bailers made of all these different types of material. But at the end of his career, contaminants knows as PFAS became an important concern. Per- and polyfluoroalkyl substances (PFAS) are a class of manufactured compounds that are extensively used to make everyday items more resistant to stains, grease, and water. These chemicals have been used in a variety of industrial, commercial, and consumer products. However, some of these products are also in sampling equipment, clothing, and other materials that may be brought into a sampling event.

With the advent of PFAS chemicals identified as a concern, sampling protocols had to be changed, starting with the types of materials used for sampling and the clothing worn by the samplers because use of these products could potentially contaminate samples during sample collection. With a relatively high probability of PFAS cross-contamination, care needed to be taken and changes were made to sampling equipment and the clothing worn during sampling.

During the week, Nick had been meticulously alternating the downrange crew to avoid heat stress, giving them a chance to recover from the extreme heat and loss of body fluids from the protective clothing. The crew gathered around the well casing after removing the lock and cap and were watching to see what was in the first bailer as the bailing line was being pulled up in the rhythmic motion of an experienced crew. Nick was just thinking how glad he was that they were in Level B supplied air protection after seeing the response on their air-monitoring equipment when the well cap was removed. The needle had pegged, and the alarm was going off. As the bailer was being pulled up, he was saying to the crew how careful they needed to be filling the sample containers without getting the well water on them. Just as he was finishing the sentence, an audible gasp migrated from the masks of the sampling crew. As the bailer was pulled, a rainbow of colors came into view in distinct layers in the bailer—reds and greens and purples and blues. Just as quickly, almost instantaneously as if by magic, one of the layers solidified in the bailer as it hit the air. Staring at the bailer, and then each other, someone in the crew just blurted out, "Air polymerization. Holy cow! What is in this shit?" The liquid in one section of the bailer polymerized as the dumbfounded crew just looked at one another each searching their memories of the inorganic and organic chemistry lessons they had had in college.

Interfacial polymerization happens when a chemical reaction is confined at the liquid–liquid or liquid–air interface, in which many small molecules (monomers) join together to form a large molecule (polymer). They were lucky it did not blow up, because often the reaction produces heat and pressure. Industry professionals carry out these processes under closely monitored conditions. Vigorous polymerization can be extremely hazardous; once started, the reaction is accelerated by the heat that it produces. The uncontrolled buildup of heat and pressure can cause a fire or an explosion or can rupture closed containers. Luckily, none of that happened during this event, but it certainly startled the crew.

Although the first thought was air, the group had no idea what caused the reaction. It could have been the temperature change, sunlight, ultraviolet (UV) radiation, or something else that triggered the reaction. The unknown mixture may have undergone polymerization just because of the slight heat change as the bailer was pulled to the surface, or it may have been the exposed to light or maybe even slightly mixed with the layer above as the bailer full of sample was rhythmically retrieved. Certainly, the types of chemicals that can cause such a reaction existed at this site because they are common from petrochemicals. It was the next day when Nick suffered his serious heat stress episode. His mistakes almost cost him his life, and he definitely learned a hard-earned lesson from his dumb mistake.

One lasting memory of this site work—and certainly this site—was Gatorade. They brought four large round coolers, the kind you see on the sidelines of football games, and filling them with ice and water and mixing in the packages of dry Gatorade at the hotel each morning was a daily ritual. They often alternated flavors as they mixed it in an attempt to make it drinkable. Later in his adulthood, Nick was still unable to drink Gatorade even though he would buy lots of it for his son during his travel and high school soccer travel years. By his son's time it was many Gatorade generations later, and the drink was sold in premade plastic bottles in a wide variety of exotic flavors. It dd not matter; Nick never could drink it again because of the endless days of pushing Gatorade down to keep his electrolytes right. It was not the only thing that caused him flashbacks from his years in those hazardous encounters.

A special thing happened to Nick and the crew after their work, something that had never happened before and never happened after despite all the arduous work he was part of. Nick and the Buffalo crew were praised by the lead EPA engineer in charge with a formal letter sent to the E&E higher ups. The letter read as follows: "I would like to express my appreciation to you and especially the following members of your staff who assisted in the recent sampling effort on the Petro Processor's site in Baton Rouge, Louisiana." It listed Nick first along with the rest of the crew. The letter went on to state: "Each performed their duties in a very professional manner, oftentimes under very long hours and intense situations. Thanks to their efforts, the project ran smoothly, and our objectives were accomplished. I look forward to working with your staff in the future." The letter was signed and dated November 1, 1983.

Years later, after Nick's site visit, all contaminated source areas on the site were capped by the early 1990s and the bayou was rerouted away from the impacted area. Other remedial actions included the installation and operation of a full-scale groundwater treatment facility along with maintenance and monitoring activities. The treatment system utilized an incinerator to treat liquid organics and air strippers for toxic fumes. The total cost of the remedial action was reported to be approximately $32,827,799.

Chapter 13

LOVE CANAL AND THE PFOHL BROTHERS LANDFILL

"Have a nice day, Jedidiah," she called as I was leaving the gym on this cold and blustery February Buffalo day. I moved to Buffalo in an attempt to get a feel for the life of Nick McCarthy and a sense of the area where he worked and played most of his adult life. One of the first things I did was join one of the oldest gyms in Buffalo, the Buffalo Athletic Club (BAC). I joined this club because during Nick's time he had a lifetime membership at the Eastern Hills Racquetball Club, which later became the BCA.

Although the BCA had morphed into different gyms with different names since Nick's time, the old name and feel was brought back out of nostalgia. It's funny how things stay the same even after so much time. I liked this gym for a few reasons; one was that it did not endlessly play loud mind-numbing techno music, which would have made me crazy within minutes, and, two, because it was close with the added benefit that it had an affiliation with my gym back in Cortlandt.

Usually, my first move in the gym was to go directly into the sauna fully dressed in my work-out gear to get the cold out of my old joints and then go on to my workout warm and ready with a sweat already starting. After a few months, I noticed the same people working out, and their habits matched mine in routine and actions. Occasionally, I went at other times and noticed a different set of people. I was determined to keep this old body in shape for the early kayaking, which I knew was still at least 3 months away in Buffalo.

Part of my training was to observe people. The one thing that became clear to me having observed so many different people in many different places was that humans are tribal. It's baked or hardwired into our DNA. It's survival instinct. It's fight or flight. People bond on all sorts of levels out of necessity or habit or both, and they associate on both macro and micro levels like the bonding of the set of people who see each other on the same days at the same times in the gym. Humans bond through various levels of association, strong or weak, which are typically based on some need for sameness and safety such as gender, sexual orientation, race, religion, community, country, look, shape, size, and so on. We chant together for our high school team, our professional sports teams, our geographic region, our college, country, company, or some other sameness. According to Merriam-Webster, the herd instinct is "an inherent tendency to congregate or to react in unison."

After my workout, I rushed out into the cold February evening almost running to get into my truck and away from the biting cold. When I took this assignment, I knew I would be in Buffalo for at least two years, so I rented a really nice apartment knowing that I was not going to give up my beautiful house in the magnificent Hudson Highlands overlooking the Hudson River. I took this assignment after my retirement for the extra funds to finish paying for the summer cottage I was building tucked away on a beautiful trout stream near Roscoe, New York, in the Catskills. As much as I wanted to leave the obnoxiously run progressive New York State, I could not bear to leave the magnificent nature of the area to which I felt spiritually connected.

Buffalo is more Midwest than downstate New York. It's not that people in New York City or its surrounding area are not friendly—no, far from it. It's that there is a certain aggressiveness to New York City area people in both speech and attitude. As had Nick, I had to learned to adjust my style and pace. The city of "Good Neighbors," as Buffalo was called, truly had a warm and convivial population. Despite my short time there, I made some great friends in Buffalo, and it was common after work to find me at one of my favorite haunts sipping on a bourbon. One place I had discovered, perhaps by chance, was a place owned during Nick's era by his childhood friend Billy Metzger. Gene McCarthys and the Old First Ward (OFW) Brewing Company was located in Buffalo's old first ward: a combination of South Buffalo Irish neighborhood nestled among grain elevators, old urban family homes, and a network of railroad tracks.

After completion of the Erie Canal in the early 1800s, many Irish workers who built "Clinton's Ditch" settled the First Ward and worked in the nearby grain elevators and manned the many Lake Erie freighters that were docked only blocks away. Founded in the late 1800s and early 1900s as the Hamburg Street Tavern, the place is still a favorite of workers at the nearby businesses, beer enthusiasts, and people young and old. One of the best places for fish fries, beef on weck, and corned beef and cabbage. Nick had lost touch with Bill in high school, not knowing at the time that he had moved away. It was certainly fortuitous that he should run into him again in Buffalo at a place that became one of his and now my favorite bars.

My other two favorite places to hang out in Buffalo are the Gypsy Parlor on Grant Street on the West Side or at my friend Marty D's, The Roost, also on the West Side on Niagara Street. I found the Gypsy Parlor by chance on a recommendation from a woman I had met. The Gypsy Parlor is located in the heart of the West Side of Buffalo. A nonconformist bohemian place for sure and any night you might find tango or trivia, drag shows, belly dancing, jazz, DJ acts, live music, and open mic nights. it's a special place, and I always have great conversations with the patrons like Eve, the owner Gabby, and the Parlor's special group of bartenders from Felicia to Tre and Jon and others like Will, Bob and Shane. It's just an eclectic place; the diversity of the clientele and workers is unmatched in any bar I have been to in Buffalo. Marty D is a top chef, and his restaurant, The Roost is magnificent. I have never had a bad meal, and its always special.

I was winding down my research and was about to focus on some of the more notorious sites and others that Nick worked on in the Buffalo area in the early days of the industry before transitioning to his later career some 40-plus years later.

One such site was the Pfohl Brothers Landfill in Cheektowaga, New York, just down the road from Nick's office. It was once nothing more than a pig farm, or so he had read somewhere. The first time Nick became cognizant that the Pfohl Brothers Landfill was a hazardous waste site was in the winter of 1982. He had been driving past this property for over a year on his way home from work. The E&E headquarters office was located on Holtz Drive adjacent to and east of the Buffalo Niagara International Airport. When Nick left work, he would turn left onto Holtz Road and then right at the intersection on Holtz Drive and Aero

Drive in Cheektowaga, New York. Proceeding east on Aero Drive, Nick would pass the landfill, which was nothing more than irregular terrain full of shrubs, a few trees, and wild grasses with lots of phragmites that spread out on both sides of the road. He could make out a few mounded areas but never paid much mind as he traversed the short part of Aero Drive towards its intersection with Transit Road—a north/south road that divided Williamsville/East Amherst and Clarence New York.

E&E had been tasked with completing a preliminary assessment of the property, and Nick was asked to help complete an initial land survey of the property. The assignment did not make much sense to Nick since there was at least a half foot of snow on the ground. Nick's job was to hold the rod for the surveyor. They spent most of the day walking up and down snow-covered mounds and around the few trees. At one point, Nick wound up hip deep in a snowbank. The survey took most of the day, and because of the snow cover, Nick walked blindly across the site below.

The next time Nick saw the landfill was in 1987–1988 while he was working for the firm Camp Dresser & McKee (CDM). He had left E&E in late 1984 because he wanted a bigger challenge. At E&E he had become a dependable experienced field person and as such his field assignments increased. Continuing to fill up sample jars with soil or groundwater or wade shin deep in a creek to collect sediment samples or hold a survey rod for a surveyor seemed like a waste of his degree and experience.

There was a saying at that time with E&E corporate staff as they witnessed numerous regional people brought into supervisory roles in the corporate office. The saying was "As you come, so shall you go." The thing is, Nick had gotten tremendous experience because when sent out in the field he and his team ran things, made all the decisions, and more importantly learned the nuances of what it takes in labor and equipment to complete various field jobs.

It was experience he probably would not get anywhere else; he was dropped off in a foreign place and succeeded. He came to realize the gift he got working at E&E. But he also knew it was his time to leave and progress. He did of course have an advanced degree in Public Health, and it was time for him to advance his career. Little did he know that he would be filling up sample jars the rest of his career and running field jobs, especially when he started his own firm and had to be the chief cook and bottle washer to keep the firm afloat.

Nick was hired as the regional Health and Safety Supervisor and lead field manager for the CDM New York/New Jersey regional office. He was hired primarily to work on projects under the EPA remedial investigation/field investigation team (REM-FIT). This was the evolution contract from the FIT contract Nick worked on at E&E. CDM maintained a small Niagara Falls office, and although Nick was based in the Edison, New Jersey, office, this small Niagara Falls office was tied to the regional CDM grouping. Because of Nick's experience in Buffalo, they loaned him out to work on the proposal for the New York State Department of Environmental Conservation (DEC) to investigate and design a remediation for the Pfohl Brothers Landfill in Cheektowaga, New York, the same landfill that Nick had walked across holding a survey rod five years before. This was the type of assignment Nick had taken the job for. No longer relegated to traveling from one hazardous waste site to another, he was now expected to write a complex winning proposal to investigate and remediate a major hazardous waste site.

The name Cheektowaga comes from the Erie-Seneca word, Ji-ik-do-wah-gah, or "place of the crabapple tree." This was the hunting and fishing grounds of the Seneca tribe of the Six Nations of the Iroquois Confederacy. The other tribes were the Onondaga, Cayuga, Oneida, Mohawk, and Tuscarora. The Tuscarora tribe had been exiled by North Carolina colonists and moved into New York to Ohagi (Crowding the Bank) near Piffard, joining the other tribes and making it the Six Nations.

The Seneca took the land from the tribe called Neutrals who were called Attawandaron, meaning "people whose speech is awry" or "a little different." One of the main villages was reportedly located in central Cheektowaga and called Ga-sko-sa-da, or Falls Village. This village reportedly stretched along an Indian trail on the north bank of the Cayuga Creek and long houses were built from Borden Road, near Broadway, along the creek bluff to Union Road.

With the exception of the Oneida, the tribes joined the British, and because of the outcome, the American Revolution proved disastrous for the Iroquois. The tribes defeated by the Americans were forced onto reservations. In particular, the Seneca were slaughtered and banished from their ancestral home in the Genesee Valley to the Buffalo Creek Indian Reservation, which was located across central Erie County.

Nick felt the hot, muggy, Buffalo morning as he left the airport terminal and proceeded to the rental car office. It often got warm in Buffalo in July, but typically it was not muggy—at least not as muggy as it got in Jersey. He was meeting the rest of the crew who had been onsite for less than a week, and Nick was coming up for his first look at the project they had won earlier that year. The project had been delayed from early spring to mid-summer, and the property was now overgrown with wild grasses, shrubs, and phragmites. In order to do the investigation, CDM had to hire a local firm to cut transects every fifty feet on both sides of the road. The transects had been completed late the day before he arrived.

Nick met Lee at the site, and the look on her face told him instantly something was wrong. "What's the matter, Lee?" Nick asked as he gave her a hug. Nick had worked with Lee on a number of jobs, and they had become friends. She was the project manager assigned to this site. "Oh my god, Nick," she said with a look of dismay. "There are hundreds of drums on this site, Nick, and I have only been down a few transects. I wanted to wait for you before doing more because I am concerned about our safety!" she said with even more stress in her voice. "Wow, I had no idea when I wrote the proposal," Nick replied. "There was no documentation suggesting that number, and I never saw them when I was here years ago because the site was under a half foot of snow," Nick continued as he unpacked the photoionization meter that he brought to monitor organic vapors. After the detector warmed up, Lee and Nick walked down random transects on either side of the road. To their amazement, as they traversed each random transect, they encountered drums—lots of drums—sticking out in all different directions, some still full and many more only partially full or long empty.

The site was described in later EPA documents as follows: "The Pfohl Brothers Landfill site was a 130-acre area that was an inactive landfill located in the Town of Cheektowaga, Erie County, New York. It was located in a mixed commercial/residential area of the Town and was approximately one mile northeast of Buffalo Niagara International Airport. The site was bordered by wetlands, Aero Lake, Aero Creek, and the New York State Thruway to the north. The remaining boundaries consist of Transit Road to the east, a Niagara Mohawk Power easement, and wetlands to the west, and residential yards (along the north side of Pfohl Road) and Conrail tracks to the south. In addition, the site was bisected by Aero Drive.

"Aero Lake, a 40-acre, twenty-foot-deep man-made lake, was created from a borrow pit used during the construction of the Thruway. The Lake was used by local residents for fishing in the warmer months. The surface drainage in the area was generally to Aero Creek, Aero Lake, adjacent wetlands, and unnamed tributaries, which eventually drained into Ellicott Creek. Ellicott Creek empties into the Niagara River at the City of Tonawanda. Land use in the vicinity of the site consists of a mix of residential, commercial, and industrial properties. Landfilling operations at the site were conducted from 1932 to 1971. The landfill was operated as a cut and fill operation; waste and drums which were filled with substances that could be spilled out were emptied into shallow 150-foot diameter pits and consisted of municipal and industrial wastes. Steel and metal manufacturers, chemical and petroleum companies, utilities, and manufacturers of optical and furnace-related materials were among those firms whose wastes were reportedly disposed of in the pits."

Besides the myriad drums and waste associated with them, initial investigations of soil, sediment, surface water, and groundwater showed the presence of benzene, chlorinated benzenes, and nitrogen compounds as well as elevated levels of polycyclic aromatic hydrocarbons (PAHs), phenols, barium, lead, chromium, cadmium, and nickel. Up until 1983 and into 1985, adjacent residents were using groundwater as their drinking water source, and connecting them to public water supply was one of the first actions.

The remedial investigation/feasibility study (RI/FS) that Nick proposed was initiated in 1988, and it confirmed significant contaminants of concern included in the various media. The RI/FS also concluded that the site posed a high potential for impacts to terrestrial and aquatic wildlife and resultant degradation of these critical environmental areas, including the freshwater wetlands, fishing areas, and creeks, as well as the uncovered and exposed waste at the site. In 1994, the site was included on the NPL.

A final report that summarized the final remedial actions that started in March 2001 stated the following: "Almost 5,000 waste-filled or empty drums were removed from the site; fifteen of which contained low-level radioactive waste. About 540,000 cubic yards of waste located along the roads were excavated and consolidated on the interior portions and capped making the area that fronted the road available for redevelopment on either side of Aero Drive and along Pfohl Road. An additional 9,200 cubic yards of contaminated soil and waste were excavated to protect the wetlands and consolidated on the interior portions of the site.

The excavated areas were backfilled with clean fill and topsoil and were reseeded.

Two caps totaling 94 acres were constructed over the consolidated wastes. Each cap consists of a six-inch gas venting layer overlain by a layer of filter fabric, a 40-mil thick very flexible polyethylene (VFPE) liner, a 24-inch barrier protection layer of clean soil and topped with six inches of topsoil capable of supporting vegetation. Forty-nine gas vents were installed to convey the gas from beneath the low permeability layer of the caps via the gas venting layer to the atmosphere. A leachate collection system was installed consisting of an eight-inch diameter perforated collection pipe set in a granular material-filled trench, which runs along the 10,000-foot perimeter of the landfilled areas at a depth of approximately five to 22 feet below ground surface (bgs.)

All of the collected leachate was discharged directly to the Buffalo Sewer Authority's Treatment Plant via the Town of Cheektowaga's sewer system. All disturbed areas of the site were subsequently restored. A vegetative layer consisting of hardy, shallow rooted grasses was established on the surface of the landfill caps. The completion of the implementation of the site remedies was issued by EPA on December 10, 2007. The site was deleted from the National Priorities List effective September 22, 2008."

Speaking in riddles that make no sense, a vice president of the United States actually said the following word salad: "It is time for us to do what we have been doing. And that time is every day. Every day it is time for us to agree that there are things and tools that are available to us to slow this thing down." Is it any wonder that no one took her seriously?

Riddle speaking was certainly not uncommon in the hazardous waste industry, and a word salad of government acronyms made it impossible for anyone but an industry insider to make sense out of what was being said. The hyperbole and conflation that is typical in the new media did not make things any better and were often used to scare the public, sometimes necessarily and sometimes not. It was Nick's profession's job to try to know when it was necessary and to convey the information into understandable information for the general public. It was a gift that not every scientist and engineer had.

One the things Nick always appreciated about E&E was that they hired technical editors whose job it was to make what the professionals wrote understandable to the public while retaining the correct amount of technical information to maintain credibility in the scientific and engineering communities. A lot of the scientists and engineers took great umbrage that editors would change their words. Nick always welcomed it, because he knew writing skills were not his strength, but the science was. He always felt that as long as the technical information was not changed, anyone one who could make his stuff more readable was more than welcome, and he figured if he paid attention to the edits, he could eventually become a better writer. Nick's main issue, besides not being able to spell and having a severe case of dyslexia, was a problem that many scientists face; they always want to add one or two more adjectives to explain something in more detail.

• • •

One of the more famous sites Nick worked on—although superficially—was Love Canal.

Nick walked into the Love Canal EPA field office in Niagara Falls, New York. He was working on another site not far away putting in bedrock groundwater monitoring wells just upgradient of Devils Hole in Niagara Falls, New York. He went to see his housemate Mike, who was running the office as the E&E staff member for the EPA. All of a sudden in walks Lois Gibbs, a primary organizer of the Love Canal Homeowners Association. She was very friendly and rather congenial as this was Nick's first opportunity meeting her as she exchanged pleasantries with the EPA staff and Nick.

The encounter left a lasting memory, but not the expected one. It was how Lois's demeaner changed when the local news media rolled in with their camaras and lights blaring. She put on a grievance show, becoming very loud and somewhat obnoxious. She was working for a just cause, and the exigent demand of her role required a certain response, but something about that exchange stuck with him— and not in a good way. He would never meet her again. Much later in his career, around 2012, his path crossed briefly with Erin Brockovich, the other famous environmental activist of his time. Both women became active chasing down the culprits of chemical releases to the environment that caused devastation to their communities caused through very poor human behavior and choices. Brockovich was famous for the movie that documented her fight in 1996 that linked a handful of cancer cases in California to contaminated drinking water.

Nick was contacted by one of his acquaintances about the medical mystery in a town called LeRoy near Batavia, New York, where a group of teenagers were displaying symptoms similar to Tourette syndrome. Nick had been to Leroy a few times because his friend and housemate Mike grew up there. It was not far from Buffalo, just down Route 5 from where he lived, and he had gone to a famous steak house in that town a few times. He also often passed through this town many times as an alternate route to the New York Thruway when he had time and wanted to take the "backroads" to Rochester.

He was told that Erin Brockovich had a theory that the two-dozen people, mostly teens but also one 36-year-old, were experiencing uncontrollable tics, seizures, and outbursts that might have been caused by a chemical spill in the town more than forty years earlier. Her theory was wrapped around thinking that these people were somehow exposed to a plume of chemicals in the environment from a 1970 train derailment in LeRoy, which dumped cyanide and trichloroethylene (TCE)—a chlorinated hydrocarbon used to de-grease metal parts—within three miles of the village's high school. Medical science on TCE shows that it can affect the central nervous system and cause dizziness, headache, sleepiness, nausea, confusion, blurred vision, and facial numbness. She was connecting this as a possible cause to the symptoms among LeRoy's local teens.

Nick learned his lessons in public health well, and these were buttressed by his many years of experience in environmental assessment and remedial design. He knew well the basic principles of toxicity and exposure; there needs to be a pathway. Something can be highly toxic but not be a hazard; without a pathway, there is no threat because there is no chance of exposure. While not completely dismissing Ms. Brockovich's theory, there were a few red flags that jumped out at Nick pretty quickly. Nick did some preliminary pathway analysis; there was enough to suspect that the train chemical release and the symptoms were not related because there was no apparent pathway. Also, the symptoms of these students did not line up with the typical reaction to exposure to these chemicals that was fairly well documented in science.

His initial thought was simply that the high school and its students were likely not being affected by this spill because no pathway existed. Exposure to the chemicals from this spill could only come in two pathways; from drinking the contaminated groundwater or by breathing the vapors that would migrate from the plume, causing an indoor air issue. With no ingestion or inhalation pathway, there could be no exposure. The kids were not drinking the groundwater, and the groundwater was flowing in the opposite direction. Therefore, a plume affecting the school was not likely being caused by this event, which meant vapors could not be entering the school from beneath the slab.

He provided his basic first blush analysis to his acquaintance; however, a role for him in the issue never materialized, and so he never had any further involvement and never had the opportunity to meet Brockovich or look further into the mystery. The teens who previously had never exhibited other symptoms of a neurological deficit suffered these tics and other neurological symptoms that seemed to mimic Tourette syndrome. Most of them eventually reportedly recovered, but the mystery lingers with various theories, including adverse reaction from immunization, most of which were easily disputed using correct scientific principles. It would not be the last time that fake science would be used to advance some benefit to those advancing their absurdly inappropriate malapropic science as Nick later discovered during the Pandemic of 2020. Nick loved trying to solve these types of mysteries. Because of his background in public health and his experience in environmental contamination, it was not the first or last time he would use his expertise in environmental forensics to solve either a human or environmental health issue.

One of Nick's favorite things to do when he worked on projects in that section of Niagara Falls was to walk down into Devil's Hole. The Niagara Gorge begins at the base of Niagara Falls and ends downriver at the edge of the geological formation known as the Niagara Escarpment near Queenston, Ontario, where the falls originated about 12,500 years ago. The Niagara River drains into both Lake Erie and Lake Ontario, and the Falls has the highest flow rate of any waterfall in North America. More than six million cubic feet of water goes over the crest every minute. Starting in Devil's Hole State Park, Nick would be awestruck as he would descend some 200 vertical feet in a little more than a quarter mile through the magnificent scenery and exceptional beauty along the walkway that leads down from the upper level of Devil's Hole State Park to the base of the Niagara Gorge and the explosive rapids in this part of the Niagara River.

Historically, it was the location of the portage route used by Native Americans to bypass the rapids of the Niagara River and Niagara Falls. Once down at the bottom, he would stare uncontrollably at the immensity of the extreme current of more than 20 knots and the presence of underwater rocks, some jutting out here and there as the water whirled by at breakneck speed. The water undulates, the height of the waves from trough to crest is approximately 10 to 13 feet under normal conditions with the bifurcation changing chaotically in its turbulent flow around large rocks here and there. These powerful rapids, the result of all the power of the water flowing over the falls above being compressed and forced through the relatively narrow river channel at the floor of the gorge, were awe-inspiring and well worth the visit if you are in shape.

Nick climbed through along the walkway with little knowledge of the history that he would only learn later. One hundred British soldiers were killed in the attacks directed at the convoy of wagons passing through Devil's Hole on September 14, 1763, as it was met by hundreds of Seneca Indians waiting in ambush. The Indians held this place both spiritually and physically as their portage location around the Falls. Poised to do close range battle that they were superior at and waiting in ambush, only three of the British soldiers were able to escape. Nick knew little of this history; when he reached the bottom and the water's edge, he surveyed a few fisherman along the wet, slippery rocks at the edge of the raging river. The power was just immediately impressive as waves crashed by at incredible speeds. Then there is the amazing sight of the whirlpool in the distance caused by the entire Niagara River being forced to spin its way through a sharp bend in the gorge, creating an amazing and dynamic sight when viewed in an aerial position.

The walk back up the gorge is not for the feeble or unfit; the ascent is moderately steep and long. The path is an impressively crafted stone staircase built into the edge of the cliff. Luckily for Nick, he was not long removed from his college athletic days and still youthful, not yet afflicted with the ailments of old age. Local literature describes the hiking trail as beginning in the wooded gorge and at the bottom offering an up-close, spectacular view of the gorge's rapids. It goes on to describe how it is one of the most popular trails for hiking and for fishing but warns that "extreme caution" must be used by following the signage and staying on the trail—and of course staying off the slippery rocks at the water's edge.

A summary background on Love Canal from references reads as follows: "The Love Canal Superfund site is located less than a mile from the Niagara River in Niagara Falls, New York. The 70-acre area includes a 16-acre former industrial landfill. In the 1890s, William Love dug the canal for a hydroelectric project. Hooker Chemicals and Plastics (now Occidental Chemical Corporation) bought the canal in 1942. For more than ten years, the company disposed of hazardous waste at the site. It then covered the landfill. The Niagara Falls Board of Education purchased the site property from Hooker Chemicals and Plastics.

Beginning in the 1970s, local residents noticed foul odors and chemical residues and experienced increased rates of cancer and other health problems. In 1978 and 1980, President Jimmy Carter declared two states of emergency for the site and evacuated more than 900 families from their homes. The severity of the contamination led to federal legislation dealing with hazardous waste, including the passage of the Superfund law in 1980. EPA added the site to the National Priorities List (NPL) in 1980. EPA worked with New York State to clean up the site. EPA and the state completed remedy construction in 1999 and took the site off the NPL in 2004. Today, more than 260 restored homes and 10 apartment complexes are located on site. Commercial and recreational uses are also on site. They include the Cayuga Youth Athletic Association and baseball fields. The site also includes a creek and a wetland. Vacant properties nearby are available for commercial and industrial redevelopment."

Chapter 14

VIBROCORING ON ALCYON LAKE AND SAMPLING RABBIT RUN

I took a short trip back downstate to go on a few kayak trips with some of my friends, Barry and Kevin, prior to completing more research on Nick. We went to a park that I knew from my work was a favorite of Nick's because his oldest brother John would take him fishing there for yellow perch and he had gone there a few times with his high school girlfriend.

Did you ever take a walk in the woods that you had never been in before or maybe you walked down a particular path many times before but never really paid attention to your surroundings with any detailed rigor? All of a sudden you notice a particular tree along the path; a unique tree with a peculiar crook. As I thought about our planned kayak trip to one of the lakes in this park, I recalled such a walk there, and my focus was on the nature around me I had never bothered to concentrate on before. As I thought back on this one particular hike, I recalled focusing on something I had seen many times before but never really noticed with any detail because it was there in my past views without distinction on my part.

At the base of this particular tree was a large rock, not quite a boulder, and on that rock was some moss. I recalled that the air along the path was slightly cool and breathed easy. The shadows from the thick adjacent woods moved across the rock as clouds passed across the sun, only visible because the tree canopy was less intense at this location. But it wasn't the moss that really caught my eye as my mind recovered this previously insignificant detail. Of course I did recall the moss being there too, perhaps because it was seductive moss (Entodon seductrix), which has a very immodest appearance and is the type you certainly would have noticed before. No, I honed in on—almost as if it was for the first time—the gray and blue lichen growing close cropped to the rock that I had seen many times before. The lichen growing on the dark gray almost black rock seemed to be part of the rock giving it an added unusual visual pattern apparent to you only when one actually stopped and really focused on the rock and its wonderful patterns.

I noticed these patterns again during the kayak trip as I hugged the shoreline in my kayak around Canopus Lake in Fahnestock State Park. My mind rejoiced at the scene, as if I was back in my beloved Hudson Valley Highlands surrounded by the topography Nick grew up with. The Hudson Highlands region as described in a post on the "Trailism" describes "the Geology of The Hudson Highlands as representing the outcrop belt of crystalline basement rocks of the northeastern extension of the Reading Prong and are a northern extension of the Ramapo Mountains in northern New Jersey."

The post further states that "to the north and east, the boundary between the Hudson Highlands, the Taconic Mountains, and the Western Connecticut Uplands region is indistinct, partly because of the occurrence of numerous thrust faults and large igneous Middle Paleozoic intrusions in the region. In general, the name Hudson Highlands refers to the low mountains or high hills that border the Hudson River north of Peekskill and south of Newburgh, New York." The rocky exposures along the trails consist dominantly of granite and amphibolite gneiss that contain quartz veins and sometimes by migmatite (a black, shiny, highly magnetic iron mineral).

The 14,086-acre park, Clarence Fahnestock Memorial State Park, includes an area in both Putnam and Dutchess Counties in New York. The original land of about 2,400 acres was donated in 1929 by Dr. Ernest Fahnestock as a memorial to his brother Clarence. Clarence was a Harvard graduate who died in the post-World War I influenza epidemic while treating patients with the disease. I recalled that Nick spent lots of time in Fahnestock as a young man fishing with his oldest brother who took him there to fish for yellow perch, and later Nick and his girlfriend would take drives there. Once they were inundated by mosquitoes when they ventured out for a walk among a beautiful grove of rhododendrons, causing them to run back to the sanctuary of the car. After the Kayak trip, my mind free, I got back to work

Lee looked over at Nick as he sat crumpled in the passenger seat on the drive northeast along the New Jersey Turnpike from the Lipari Landfill in Gloucester County towards the office in Edison. "How are you feeling?" she asked. "Terrible," Nick replied barely above audible. "My head is throbbing; I don't think I have ever had a headache this bad in my life," he said with slightly more energy. "I'm hoping the aspirin I took starts to kick in soon before my head explodes," he continued almost in rapid fire, knowing he had only so much energy to even say that. Lee and Nick had become good friends working on this site and others at

Camp Dresser & McKee (CDM), and he could tell she was genuinely concerned. Typically, the crew stayed at a hotel, but this was the end of the week, and they were heading home for the weekend.

About 15 minutes later when the aspirin kicked in and he was feeling slightly better and able to think, he realized what had happened. He had been working all day downwind of the Fruit Orchards that were immediately adjacent to the site, and they had been spraying the orchard all day. Typical of Nick, being so focused on the safety concerns from the site, he totally forgot about the effects of being exposed to the pesticides being sprayed on the orchard until he realized his symptoms were textbook pesticide exposure.

He of course should have known better, and as he was beating himself up about it still slumped in the passenger seat, he told Lee what he thought brought on his migraine-type headache. "Well," she replied as she pulled into the parking lot to drop him off, "I am just glad you are feeling better because you were white as a ghost." Lee lived in New York City and worked in the New York office, so she was off and on her way to beat the traffic. Nick, feeling better, headed home to his apartment in Highland Park, New Jersey, just across the Raritan River from New Brunswick and Rutgers University. They had been working on the Lipari landfill located in southwest New Jersey completing some groundwater well sampling and also some sampling in Rabbit Run and the associated wetlands that ran adjacent to the site.

The Lipari Landfill site off Route 322 in Pitman, Mantua Township, New Jersey, was originally a 16-acre sand and gravel pit. About a six-acre portion was eventually turned into an unregulated dump that took municipal and industrial wastes of all types. At the time Nick was onsite, it was ranked the No. 1 Superfund site in the nation. Pitman is about 20 miles southeast of Philadelphia, and some of the major industries in and around Philadelphia and Cherry Hill, New Jersey, sent their waste to the landfill. Between 1958 and 1971, household waste, liquid waste, semi-solid chemical waste, and other industrial materials were disposed into the six-acre portion of the property. These wastes included solvents, paints and thinners, formaldehyde, dust collector residues, resins, and solid press cakes from the industrial production of paints and solvents.

During 13 years of operation, some three million gallons of liquid chemical/industrial wastes and 12,000 cubic yards of solid chemical/industrial wastes were dumped at Lipari.

One explosion and two fires were documented at the site, and the landfill was closed in 1971 by the New Jersey Department of Environmental Protection (NJDEP). Contaminants seeped into the underlying aquifers and leached into the nearby waterways, including a marshlands/wetlands, Chestnut Branch, Rabbit Run, and Alcyon Lake. The lake was part of a park, and because of the contamination found by Nick and other investigators, it was subsequently closed for recreational use.

Pitman is a borough in Gloucester County, New Jersey, and was named for Rev. Charles Pitman, a Methodist minister. Nick worked on the landfill on and off for a few years; the remedial investigation included installing deep groundwater wells in and around the landfill and sampling other media, including vapor escaping the landfill/landfill gas and surface water in the wetlands and creeks. On one of his site visits, he was managing a crew that was going to sample sediments in Alcyon Lake. This required subcontracting work to a drilling firm that would use a makeshift barge with a hole in the center to use a technique called vibrocoring to retrieve sediment samples under water. Vibracoring is a technique for collecting core samples of underwater sediments and wetland soils.

The method uses a hydraulic or pneumatic mechanism called a "vibrahead" attached to a core tube, which is driven into sediment by the force of gravity, enhanced by vibration energy. When the insertion is completed, the vibracorer is turned off, and the tube is withdrawn with the aid of hoist equipment. Vibrocoring is an effective method for obtaining samples from thick sediment deposits. It produces a relatively undisturbed core from which samples are collected for laboratory analysis. The plan was to maneuver the barge across the lake in a grid pattern, and samples would be collected at selected grid locations to get representative sediment samples across the Lake. The lab data would be used to characterize the lake sediment in terms of site contaminants and to design remedial techniques depending on where the sediments where impacted, at what depth, and how badly.

It was during the sampling at Alcyon Lake that Nick witnessed something he had read about in environmental journals of the time but had never seen. Alcyon Lake was one of those typical suburban/country parks wrapped around a small lake. It had swing sets and picnic tables and geese and ducks that parents and children fed. Nick of course by this time had heard the horror stories about six-pack beer rings being a hazard to wildlife. Most of the work they were doing was at the far end of the lake away from the public area.

One day they decided to take lunch across the lake at the picnic tables. As they were sitting talking, up waddles a group of ducks looking to be fed. One particular duck stood out because it was adorned with a beer ring around its neck. The duck had somehow got its neck through one of the six pack rings and was wearing the plastic proudly as it waddled around the shore on the lake towards their table looking for a hand-out. Nick and the crew spent an hour trying to catch the duck to cut his "necktie" and free him of the hazard it posed. When they were finally successful, the duck waddled off almost in a huff at losing its adornment. The crew took to calling him Miller from then on.

After the initial investigations, the EPA completed a few Interim Remedial Measures (IRMs) at the Lipari site prior to the final remedial measures. IRMs are typically completed early because some immediate hazard is determined that may be creating an imminent threat to the public or the environment. Earlier remedial measures included the installation of a liner cover and slurry wall enclosing the landfill and a leachate treatment system. The persistence of odors following these measures led the EPA to conclude that further remediation, this time in offsite areas, was also necessary.

An environmental remediation firm based in Niagara Falls, New York, was contracted to excavate and treat contaminated soils and sediments from Alcyon Lake and associated marsh and stream areas adjacent to the Lipari Landfill and restore these areas to natural conditions. All work was conducted under U.S. Army Corps of Engineers and USEPA Region II oversight. There were many firms associated with investigation and remedial activities at Lipari over many years, and it was this site along with his experience at many others that made Nick realize how slow to act government is. In Nick's opinion, even with the early IRMs at Lipari, these sites were "studied to death" before any real meaningful remediation was completed, and government scientists often turned these sites into research projects before moving towards meaningful remediation. Perhaps that was a product of the early days of the industry. To Nick, it was a clear sign that it was people who had no business motivation who were constantly funded.

When meaningful remediation was finally started by this firm, it included three task areas as defined in public site documents:

1) Site preparation, which included installation of 1.5 miles of double-wall HDPE force main, complete with electronic leak detection; constructing over 2 miles of temporary haul roads, including stream crossings; diversion of Chestnut Branch; and construction of soil erosion and sediment control measures over a 25-acre area;

2) Excavation of 60,000 cubic yards of volatile organic-compound-contaminated

soils from the 14-acre marsh areas using specialty equipment. These soils were treated using low temperature thermal desorption then placed in an onsite cell. The wetland areas were restored to natural conditions with native marsh vegetation, and a drain bordering the marsh was installed to protect offsite areas from recontamination from the adjacent Lipari Landfill; and

3) Installation of a surface water bypass channel, diversion berms, and floating haul roads on the exposed Alcyon Lake surfaces, followed by excavation of 85,000 cubic yards of contaminated lake sediments. The lake was reconstructed to depths and contours in accordance with United States Core of Engineers (USACE) guidance and specifications.

To protect the nearby residents, workers, and site works, a number of health and safety measures were employed since the active remedial actions would produce dust and vapors, including volatile and semi-volatile organic compounds and inorganic metal wastes. These measures included: continuous real-time air monitoring and site perimeter monitoring for the protection of the onsite workers and neighboring persons and residences; specific personnel protective equipment (PPE) and clothing when working in the hazard zones, ranging from Level D (construction gear) to Level B (self-contained breathing apparatus and chemically protective clothing) protection. The perimeter air monitoring system was set up to continuously monitor the output from the thermal treatment system and to ensure no offsite migration of VOCs and semi-VOCs downwind during excavation and staging operations.

As a contractor to EPA and the NJDEP, much of the work Nick was involved with at the site was associated with the initial actions, which included the installation and sampling of sixteen groundwater monitoring wells to determine the direction of groundwater flow and the extent of contamination; the installation of a security fence to restrict access to the landfill and neighboring impacted wetlands areas; and the sampling of soil, groundwater, surface water, sediment, and air.

After some initial IRM in 1982, EPA contractors completed the cut-off wall and landfill cap in 1984. In 2011, an additional slurry wall was installed to contain contamination that had been located outside of the original wall and also to reduce the amount of groundwater that required treatment. Additionally, a new cap was placed over the area in 2012, and other remedial actions continued well into 2017 and beyond, including the installation of a vapor extraction and treatment system and the continuation of groundwater flushing and extraction. An aerobic environment was maintained in the landfill to allow naturally occurring microorganisms to degrade Bis(chloroethyl)ether (BCEE) the site's primary contaminant of concern (COC) and to reduce the buildup of potentially explosive methane concentrations. In 2019, the NJDEP took over the daily operation and maintenance of the remedial action at the Lipari Landfill site, and in 2020, it established a Classification Exception Area/Well Restriction Area (CEA/WRA) for contamination in groundwater associated with this site. Groundwater, surface water, and air were to be monitored on a regular basis to ensure that the surrounding community and environment were not exposed to hazardous chemicals.

One of the other more significant sites Nick worked on in New Jersey was the American Cyanamid Superfund site. A summary of the property provided by the NJDEP included the following information: "The American Cyanamid Superfund site was located in an industrial/commercial area of Bridgewater Township, New Jersey. It was a 435-acre site bounded by the Raritan River to the south and west. Calco Corporation, a rubber manufacturer, began operating at the property in 1915 and the property changed ownership several times since that time. American Cyanamid purchased the facility in 1929 and expanded the plant's size to 575 acres. American Home Products purchased the site from American Cyanamid in 1994, ceased manufacturing operations in 1999, and sold the plant to Wyeth Holdings Corporation."

"Products manufactured during the plant's history included rubber chemicals, dyes, pigments, fungicides, petroleum-based products and pharmaceuticals. In 2009, Pfizer Inc. purchased Wyeth Holdings Corporation and assumed full responsibility for all environmental remediation at the site. Historical industrial activities at the plant caused areas of soil to become contaminated with volatile organic compounds (VOCs), cyanide, polychlorinated biphenyls (PCBs), and metals, and the shallow and deep ground water aquifers to become contaminated with metals and VOCs. In addition, there were sixteen surface storage units referred to as 'impoundments' that contained tars, wastewater sludges, iron oxide and general plant debris, and four hazardous waste lagoons that required closure under the Resource Conservation and Recovery Act (RCRA). The site was added to the National Priorities List of Superfund sites (NPL) in 1983."

Nick's visit to the site was focused on a sampling program across one of the lagoons off a barge that was being pulled across manually. It was one of Nick's more memorable sites even though he was there only two days because of the toxic soup that was below him in the lagoon and the stress associated with monitoring the air and keeping the lagoon fluids off people during sampling, which required strict focus.

Chapter 15

ODE TO JOHN AND THE JACKSONVILLE OIL PITS

It was an odd sight given the particular surroundings and location, and I squinted my eyes a little tighter. I had left the West Canal Marina about an hour or so before and was on my return trip paddling down the feeder stream near Old Falls Road and Lockport Avenue in North Tonawanda, New York, at its mouth with the Erie Canal/Tonawanda Creek. Tonawanda Creek makes up part of the Erie Canal from near Lockport, New York, westward to the confluence with the Niagara River. At first, I thought maybe it was a muskrat or perhaps a mink crossing the small creek. But then its motion became extreme, changing direction and speed chaotically. Also, there was the blue-like hue I thought I was seeing.

As I paddled closer and the object become more in focus, it appeared that perhaps it was a beach ball skiing on the water's surface, with the swirling wind changing its direction and speed after it had escaped from one of the neighbor's yards.

This was a favorite excursion of mine almost every time I put in at West Canal because I enjoyed the sights along this narrow feeder stream with backyards of houses ending at the water's edge. Some of the yards were manicured with flowers, fire pits, outdoor furniture. Several of them had boat docks in the canal. Other houses were wilder as if the owners paid no attention to the stream that bordered their land. I was making this trip later in the day—around 4PM—than my usual morning trips and still most of the backyards and fire pits were empty and cold as if hardly ever used. I thought often about what great parties and backyard events you could have if you owned one of these places—let alone the opportunity to have such quick access to kayaking and boating on the canal.

As I passed one yard, I was happy to see some backyard festivities ongoing up near a house a good distance upgradient from the stream. A little farther down, two young adults—a girl and a boy—were fishing, and they happily told me they had just caught a small-mouthed bass. The fish can get very large here because this part of the canal and this feeder stream were just a few miles east of the Niagara River and Lake Erie, and it was common for lake fish to migrate up the canal/creek.

When I finally got within about ten yards of the object, I realized it was a robin egg–colored balloon light enough to scurry across the water, changing direction at the whim of a slight breeze. Just as I was about half a kayak's length away, the balloon got funneled into a crevice along the shore; some downed branches captured it motionless along the shoreline, where it would be doomed to remain a mere ten yards from the canal.

I rescued this balloon because I felt it warranted a more meaningful trip and a better resting place. So, I took it with me under the small bridge that lay just yards ahead where the mouth of the feeder stream met the Tonawanda Creek/Canal. I paddled out into the middle and released the balloon with little ceremony, almost feeling like a criminal for being the perpetrator of its new life and adventure. Watching it move downstream, I began to contemplate how far this human breath would travel before it escaped back into the atmosphere. Would it even reach the Niagara River miles downstream, or would the anonymous breath be released as the balloon sored over Niagara Falls?

++++++

John looked at Nick with that typical mischievous grin as Nick was peering at him with that "fatherly" pissed-off look. John had just come out from behind the water tank, and Nick had seen him reaching down into the weeds and grabbing something. "John," Nick started, "How many times have I begged you to stop with those snakes?" John stood there with the grin that turned slightly sheepish, holding a 3-foot snake, blood dripping off the back of his hand where the snake had bitten him at least once. John started to say the well-known rhyme again with a smile: "Red on yellow kills a fellow. Red on black, friend of Jack." Nick replied, knowing it was not going to do any good, "Yeah, I know you are an expert and all, but what about getting an infection from those bites?"

He then followed up with a dig common among these two who were longtime friends after spending much time together on hazardous waste sites, "I thought you were quicker than that. How did you screw up and get bit? Too many beers last night?" "Anyway," John continued, "This is a red corn snake or red ratsnake, which is not venomous.

Nick just shook his head, saying something to John about not wanting to have to drive him to the hospital—or worse, lose him on this job and have to be a man down—and then went back to what he was doing. The day before, they had found what they thought was a coral snake in their mud pit, which caused them all to remember they were not in Buffalo anymore and they had to worry about this venomous snake along with the other hazards.

The best way to identify a coral snake is through its unique coloring. Eastern coral snakes have a pattern of banding that goes from their heads to their tails. The pattern has repeating bands that go; thick black band, thin yellow band, thick red band. As John had demonstrated, people in Florida were well-versed in the saying "red touch yellow, kills a fellow." But this is true only for North American coral snakes. Eastern coral snakes never have bands of red touching bands of black, but they occasionally have small spots or faded blotches of black on their red bands. Another "tell" for coral snakes is their inky black head, as most mimics have red heads. Non-venomous scarlet kingsnakes have red, black, and yellow (or white) rings down the body. However, the narrow yellow rings only contact the black rings, not the red rings as in the coral snake.

Now they were longtime friends who knew each other well, but when they first met, Nick did not like John. He thought he was a slightly egotistical fellow. However, as they were thrown out on sites in the middle of nowhere, it was not long before Nick got to know the smart, fun-loving, slightly goofy guy John was, with his curly brown hair and engaging smile. "Jeez, John," Nick would often say on these occasions. The game was repeated numerous times with slight variation on whatever site they were on because Nick was usually the designated site safety officer.

They were working on the Whitehouse Waste Oil Pits site in Duval County, Jacksonville, Florida. Their mission was to install six bedrock wells into the groundwater aquifer. The well locations were pre-selected in Buffalo and located around the perimeter of the site to capture both upgradient and downgradient groundwater. In this manner, it would be possible to gather a number of different pieces of site-specific data, including groundwater flow direction, hydraulic conductivity, groundwater flow, PH, dissolved oxygen, and more. Groundwater samples were also collected and analyzed for a laundry list of specific chemicals to determine the impact from the past practices on the groundwater.
Comparing upgradient water sample results with downgradient results provided

insight into the impact the site had on the environment. At each location where they picked to drill, they first dug shallow mud recycling pits and lined them with plastic to catch and recirculate the drilling fluid "mud" as it came up from the drilling operation. In the morning they always checked these pits closely before starting work, because it was in one of these that they found a highly venomous coral snake early in the project.

A few days into the work, they had decided to rent an RV with air conditioning to use as a cool place to get out of the very hot sun because for some reason they did not have a site trailer as was typical for these jobs. Additionally, they had befriended the neighbor, who told them that the area was used by kids to hang out and drink, so none of their stuff would be safe left alone overnight. This led to their brilliant idea that the crew could take turns staying the night in the RV to protect their gear and equipment. It did not dawn on them that leaving one guy alone in the backwoods near Jacksonville in the early 1980s was probably not a smart thing to do, and Nick spent a number of spooky nights in the RV alone waiting for the sun to come up and wondering what each sound was.

The RV did work great as a place to get out of the heat during the day and store equipment; the crew hung out there often taking breaks. A couple days into their work, they also decided to rent some large stadium lights and drill at night rather than during the day as a way to escape the oppressive heat. There was no way they could continue wearing coated Tyvek suits in the hot Jacksonville fall sun and complete their mission in a timely manner.

The neighbor they had befriended was a really interesting fellow. One of them, John or Nick, must have wandered over to the guy's house, which was isolated on a seemingly remote piece of property not more than 500 or so yards from the site to ask about the area and his drinking water well. The guy was super friendly and particularly interested in the goings on because he had attended a few public meetings about the site a year or two before. He invited the crew over for a few beers after work, and the guys took him up on it in the name of community relations or some such excuse. This guy had obvious physical malformations that became more evident the longer one remained in his presence.

What they learned was that he had been an Army Ranger in Vietnam who would get dropped off behind enemy lines. He survived numerous highly classified intelligence-gathering missions while serving in the darkness of the Vietnam War that had ended less than ten years before, and he had come home to a country that didn't appreciate his exploits or the war in which he fought. He had been wounded multiple times, and it showed. There were some shrapnel wounds on his face and neck as well as probably others hidden beneath his cloths. The most obvious other injuries were seen when you noticed his hands, which were missing some digits with some clearly rearranged to offer him the best dexterity.

He told the crew some interesting stories over those cold beers about how he and some of his fellow soldiers "were counter-intelligence." "A small group of us would get dropped way behind enemy lines to conduct P.O.W. snatches, check out enemy movement, and do wire taps. We tried to get a handle on what was coming south in Vietnam." He told them that sometimes he would get dropped off by himself or they would break up in smaller groups once on a mission, and they would have to make it back safely on their own. They often worked with The Montagnards, or the hill people of Vietnam, who hated the Vietnamese.

He explained how when sent on a mission he took everything he needed with him, including "a car-15" (a collapsible stock version of the M-16 rifle), an old Browning Automatic Rifle belt in which he carried thirty-six 20-round magazines of ammunition, two 30-round clips taped end-to-end in the car-15, a .357 magnum Smith & Wesson pistol, a little M-79, 40-millimeter grenade launcher, a bandolier of high explosives and phosphorous grenades, some C-4 explosives, smoke grenades, a knife, water, and food. He told the crew: "In Vietnam, you got pretty good at this after a while." He added: "You either got good or you got dead. When it's your time friends, it's your time."

This particular project was actually pretty special in terms of the work at E&E. This was one of the first times they were using the new drill rig E&E had recently purchased. The higher-ups, observing how much work they subcontracted out, thought they would keep some money in house rather than paying it out and losing that chunk of the project. One likely area that was part of every field job was drilling soil borings or installing groundwater wells. Luckily, Nick had convinced the higher-ups that they needed to hire a head driller with experience to run the rig because originally they wanted Nick and others to run the rig.

Nick had been on enough drilling jobs to know that you don't put a novice behind a drill rig. It was one of the most dangerous occupations, and so he was adamant about the experienced driller. So, they hired Matt, who had worked for the company selling this particular drill rig line, and he became part of the crew for a number of years fitting in just fine with the experienced hard-working/hard-partying corporate hazardous waste crew at E&E. The crew members were to act as driller helpers along with the technical aspects of the work. Sadly, not long after Nick left E&E, they abandoned the head driller and tragedy struck one of the geologists working on the rig. Rumors spread fast in the hazardous waste industry even as more firms entered the business, and Nick found out unceremoniously by a third party. It was additionally sad for Nick because he had this guy as an office mate for a short time before Nick left the company.

The USEPA Superfund Record of Decision and a health Assessment from the US Public Health Service/US Department of Health and Human Services summarized the site as follows: "The 7-acre Whitehouse Waste Oil Pits site was used by Allied Petroleum Products (Allied) to dispose of acidic waste oil sludges from its oil reclamation process in Whitehouse, Duval County, Florida. Allied Petroleum Products Company operated a used motor oil recycling business at the Whitehouse Waste Oil Pits site from 1958 until it went out of business in 1968."

A cypress swamp system and residential area were immediately adjacent to the site. The northeast tributary of McGirts Creek traverses the north site boundary. The Floridian surficial aquifer underlies the site and is the drinking water source for local residents. The material disposed at the site resulted when waste oil was treated with concentrated sulfuric acid, precipitating the fuel additives and sediment as well as a large portion of the metals and other contaminants in the waste oil. The acid sludge produced in the first step and clay used to decolorize the oil were dumped into the unlined pits at the site. A 200,000-gallon waste oil spill occurred in 1979 and the state and city capped the pits with clay and topsoil.

Soils and groundwater at the site were contaminated with heavy metals, primarily lead, chromium, nickel, and petroleum products, including PCBs. This site received considerable media attention. It was the subject of a 1979 ABC-TV documentary The Killing Ground and a nationally syndicated newspaper column by Jack Anderson in August 1986. Whitehouse residents, living adjacent to the site, expressed concerns at EPA/Florida DER–sponsored public meetings in 1984 and 1985.

The community of Whitehouse, Florida (population at that time of approximately 6,000) is located within 0.25 miles east and southeast of the site. The community is composed primarily of two-bedroom houses and mobile homes on one-half to one-acre lots. A low-density residential area is located west and northwest of the site, and several miles northwest of the site is the Cecil Field U.S. Naval Air Station. The area north and northeast of the site is largely undeveloped land comprising pine forests and cypress swamp.

The Whitehouse Site is underlain by a shallow aquifer system that flows southwest and a deeper Floridan aquifer system that flows south. The total thickness of the shallow aquifer system is approximately 500 feet. The total thickness of the Floridan aquifer system is greater than 2,000 feet.

The shallow aquifer system comprises undifferentiated Holocene and Pleistocene Age sediments deposited during the formation of marine terraces and beach ridges. Holocene and Pleistocene deposits primarily consist of fine- to medium-grained loose quartz sands, iron oxides, and sandy clay beds containing mollusk shell material. Underlying Pliocene and upper Miocene deposits consist of sand, shell, sandy clay, and limestone. This was the first time Nick was on a drilling job where he saw shells in the recovered soil, and it made him think about what the earth looked like all those many years ago when this was part of the ocean, probably where some beech was located with thunderous waves and coastal sea life.

Nick and the crew were experienced fieldhands by this time, but the assigned geologist was still a little raw and very much an absent-minded type. They only had a two-week window to complete the six wells, which should have been more than enough time. The plan was for the wells to be finished as open holes in the bedrock, meaning there was no well pipe or screens at the bottom. Typical groundwater well construction includes installation of the well screen, well casing, and filter/gravel pack. However, often when wells are constructed into competent bedrock, they are completed as open holes.

After construction, the process of well development must be undertaken to clean the borehole and casing of drilling fluid and fine sediments that were introduced during the drilling process. The primary objectives in well development are to establish optimal well yield that indicates the well has been adequately cleaned, which will also allow quality representative water samples to be obtained.

Upon completion of the first well, the geologist wanted to complete a pump test before drilling any other wells to acquire information to be used during completion of the remaining wells. Once the well was developed, a very expensive stainless steel Grundfos Submersible Well Pump was lowered into the well down into the open hole about a foot or two off the bottom. Within minutes of starting, the pump stopped functioning. The pump was pulled from the well and found to be clogged with rock and shell fragments. After cleaning, it was tried again only to have the pump become clogged again. Well, this happened about five or six times before Nick told the geologist that he would have to work on the pump test issues with people back in Buffalo because the crew had a schedule to keep if all the wells were to be installed by the date promised. The poor kid was somewhat devastated by the immediate failure, but Nick explained that in field work, you always had to have workarounds or the job would never get done. Left to the geologist, they would have spent a week trying to complete the pump test and never gotten any other wells installed.

The crew completed the job, and the brain trust in Buffalo figured out how to get the pump test done by installing small diameter screens around the pump intakes. As with most of these jobs, Nick and the crew never learned the outcome of their efforts before being sent off to another job.

Chapter 16

THE EGGSHELLS OF EAGLES -FROM DDT TO PFAS

The ghosts of past chemical evils and the damage they caused often played across Nick's mind as he worked on sites and contemplated events throughout his career. At these times, he often thought about his industrious ancestors who emigrated from England to start a rolling iron mill on the pristine Croton River in 1831. Throughout history, as humans advanced, wastes were part of the process as was the inevitable impacts to Mother Earth. Could the process stop as some demand? Were there truly alternates less egregious, or was the tale of unintended consequences always to play out? Nick always believed his chosen purpose was to achieve ways to reduce or fix the impacts of industry on the environment but still allow mankind to progress, which was its inevitable cause.

The alchemists of today, including the naive environmentalist hippies who get taught in "education camps" called public schools and later colleges and universities, get bamboozled by the corrupt environmental justice politicians and big corporation leaders who are no better than the snake oil salesman of the past selling present-day alternative and renewable energy "tonics" to the uneducated masses. Seedy profiteers of alternative and renewable energy exploit an unsuspecting public that no longer thinks in any form of logical progression and thus falls prey to those using them for power and control. It is frankly despicable, and its ultimate outcome has been the wrecking of the field of science as the disreputable sell phony science to the masses using a corrupt media, educational system, and big corporations pretending to care.

It all sounds good until there is no food on the table or heat in the winter, bare shelves in the market, and little workable energy to get from one place to another—the consequences of which are less felt by the carnival barkers and more by the poor, old, young, and people just living their lives day to day. The elites decide they know better than "the unclean dirty masses who certainly are not bright enough to make the correct decisions." So those masses must go without while the elites fix the world without ever feeling even a small price for their actions. It has been the same throughout time on different scales.

Nick always felt that finding "cleaner" alternate or renewable energy sources was just the right thing to do. He wanted to seek advances that were better to the planet but more so because it intrigued his scientific mind and his need to do good and be great. He really got excited thinking about ways to use wastes produced by society in creative ways either to help clean up the environment or as new energy sources in a kind of synergistic, sustainable way. Like using waste sources that when injected into soils or groundwater increase either the aerobic or anerobic conditions, thus increasing the natural bacteria's ability to use the environmental contaminants such as petroleum or chlorinated solvents as food sources, cleaning them up in the process. Or specific plant and tree species using phytoremediation as a means to process contaminants in the ground.

One such creative use of waste was using waste woodchips from trees harvested for some other purpose or waste wood to fill six or twelve inch "soxs" made of biodegradable hemp. These were then used to surround and reducing sediment leaving a construction site. Using these woodchipped-filled biodegradable sox and impregnating them with natural elements that would reduce the nitrogen or phosphate load leaving farms or metals and other pollutants leaving industrial sites was even better. Nick had thought that hydrogen fuel cells developed from waste biomass made from food waste, agricultural waste, manure, forest products, and other waste would eventually be the game changer in alternative/renewable energy when he was introduced to this concept by Denny, a colleague who was certainly a forward thinker. But as a scientist, someone who was trained in the scientific method, he was never fooled by sophomoric pitches so easily bought by the masses. He easily saw through the conflated, contorted terms meant to confuse the ignorant using nothing short of the best Madison Avenue sales pitch with blind religious overtones—the religion of green energy, some called it.

In the latter half of the nineteenth century, there was a dramatic rise in the popularity of "patent medicines" often sold on the back pages of newspapers or by traveling salesmen. These tonics promised to cure a wide variety of ailments. Similar ads were frequent on national TV commercials in the 2020s. Were some cures valid? Sure, to a degree—just as use of alternative/renewable energy makes sense in the right magnitude or when fully advanced when the science is right.

Bankrupting a society and causing great harm to humans for phony science is not noble at all!

The quiet scam is the belief that these renewable or alternative energy sources create no damage to air, water, public health, wildlife, habitats, or the land. All energy sources affect the environment in which we live, including the more commonly known renewable energy sources like solar, wind, hydro, biomass, and geothermal. There are real and measurable physical, mechanical, electrical, and environmental limitations of renewable sources of energy and both pros and cons that need to be correctly engineered and scientifically analyzed for their proper development.

Could man always think himself out of the nastiness he produced by making products and using ingredients less impactful to the world around him? Nick observed early that there always seemed to be the new bugaboo in the environmental health and safety world caused by some new and better product that sometimes eventually caused fear or distress—either out of proportion to its importance or not. Perceived or real, overblown or not, typically mankind at some point decided the benefits of this new product were outweighed by the harm. Most of these products' downfall involved one or more of its chemical constituents. When Nick was young, the most ominous was DDT.

Although developed in the 1800s, DDT was used extensively during World War II to fight malaria and typhus and became available for public use in 1945. In 1962, the publication of the book Silent Spring by Rachel Carson initiated public concerns about DDT. The book is credited with launching the modern environmentalist movement; Carson warned man-made chemicals, particularly pesticides, were a significant threat to human health. By the 2020s, other scientists believed the benefits of agrochemicals far outweigh their harm, but they provide a dearth of evidence to support this view. Actual scientists welcome scientific debate.

By 1972, DDT was banned for agricultural use. Nick was learning in his later high school years and early college years about how DDT washed into nearby creeks, streams, and rivers and accumulated in fish that then were ingested by bald eagles. This resulted in a thinning of their eggshells, which then cracked under the weight of the adult eagle incubating its young.

Other chemicals and their negative effects were also coming to light. Equally alarming to DDT was the discovery of the harmful effects of polychlorinated biphenyls (PCBs), found in insulating oil or dielectric fluid used in transformers, capacitors, and other electrical products. In 1968, over 1,200 people in Japan were poisoned by eating food mistakenly cooked in oil that was heavily contaminated with PCBs and other chemicals. This caused and illness that became known as Yusho (rice oil) disease.

There were many other products and chemicals throughout Nick's career that seemingly jumped out of nowhere as the "next greatest thing" followed by it being hailed as the "worst environmental poison." Either some spectacular occurrence created the spotlight or overwhelming evidence had accumulated that pushed the new wonder chemical or product from great to apocalyptically destructive, overriding any good it ever possessed. Just when the industry had satisfactorily responded to the new crisis in environmental health, a newer one arose; DDT, PCBs, treated lumber, urea formaldehyde foam insulation, asbestos, lead, radon, mold, and on and on. Nick recalled for example when treated wood became a concern after it was widely used in the construction industry and was part of many of the homes and playgrounds across the country.

Hexavalent chromium and pentachlorophenol (PCP) were widely used by at least the early 1970s to the early 1980s to treat wood and increase its longevity, especially when the wood would be exposed to nature. PCP, first produced in the 1930s, is an organochlorine compound used as a pesticide and a disinfectant. Registered as a pesticide on December 1, 1950, it was one of the most widely used biocides in the United States. It is a synthetic substance made from other chemicals and does not occur nature. By 1987, the use of pentachlorophenol as an herbicide, defoliant, mossicide, disinfectant, and wood preservative was halted, restricting its use to certified applications; it was no longer available to the general public. It is now used industrially as a wood preservative for power line poles, cross arms, and fence posts.

Most of the adverse effects of human exposure to PCP are really from its toxic impurities, such as polychlorinated dibenzo-p-dioxins and dibenzofuran, which are also common environmental pollutants from other pesticides/herbicides, such as Agent Orange and other more common sources. Some of the major exposures to the general population were from combustion, such as municipal waste incineration and automobile exhaust, chemical industry wastes, agricultural and industrial chemicals, and hyper-chlorination of water. It is present in cigarette ash and is a byproduct of processes in the pharmaceutical industry. At least sixty wood-preserving sites are on the National Priorities List, and in 2022 the EPA required the cancellation of PCP in any products.

During his career, Nick assessed, investigated, and remediated many chemicals that were once thought to be saviors of humans only to become major environmental concerns requiring significant legislation and procedures for their identification, assessment, removal. This fostered many cottage industries to support their investigation, removal, and substitution. One of the more notorious of these, at least in the northeast—although somewhat short lived in its anxiety-producing compared to the others—was urea formaldehyde used in foam insulation. Nick spent hours early in his career at residences to measure formaldehyde levels caused by application by inexperienced or inappropriate farms.

The other more villainous chemical and product that became chief antagonists to human health was asbestos, which was used to reduce the fire hazard in building materials and clothing. Another was lead, which had been an addition to paint to affect its pigmentation and moisture resistance. There were others, and their discoveries as uniquely harmful agents always precipitated demonstrative reactions, laws, regulations, rules, and requirements. Every time, these would give way to the next evil chemical or man-made substance. In the 2020s, PFAS was the new miscreant group of chemicals.

Per- and Polyfluoroalkyl Substances (PFAS) are a group of chemicals used to make fluoropolymer coatings and products that resist heat, oil, stains, grease, and water. Fluoropolymer coatings are blends of resins and lubricants used in products such as water-repellent clothing, furniture, adhesives, paint and varnish, food packaging, heat-resistant nonstick cooking surfaces, and insulation of electrical wires. Chemicals in the PFAS group include PFOA and PFOS, which act as surfactants used to fight certain types of fires.

Due to their widespread use and persistence in the environment, by the 2000s, most people in the United States had been exposed to PFAS chemicals. Studies were showing that these compounds remained in the environment and in the human body for many years. Because of their unique and useful properties, these manufactured chemicals have been used in industry and consumer products since the 1940s. The rapid interest in the identification of these "chemicals of emerging concern" in the 2020s resulted in the need for experts to identify them and reduce exposure to humans and the environment.

During the waning years of Nick's career, he became very knowledgeable on PFAS chemicals, and his company was at the forefront as experts in the environmental reduction and elimination of human and environmental exposure to PFAS chemicals—the so-called "emerging contaminants" or "chemicals of emerging concern," as they became known in the industry.

Nick had been working on a couple of projects impacted with chlorinated solvents and using the remedial technique of injecting specific materials to enhance the reduction of solvents in groundwater plumes and saturated soils. PFAS chemicals were also elevated at sites where chlorinated solvents were found. However, because of their inherent properties that made them such "wonder products," they were not as easy to remediate once in the environment.

For chlorinated solvents, Nick's company and team of outside experts often used microbial enhancement techniques, which fostered the natural bacterial growth using the contaminant as a food source. They often depended on the stoichiometry combined with a metallic valency such as Zerovalent iron (ZVI) to help sever chemical bonds and increase their rapid degradation. The company partnered with a university professor who had conducted research projects to find an effective means to remediate the PFAS chemicals that were often comingled with the chlorinated solvents. These represented cutting-edge remedial techniques to solve a complex contamination problem.

Just as Nick sat in meeting with experts in the early days of the industry—learning and contributing where he could—he sat with experts in his waning years. The difference in his later years was his contribution of his vast years of experience and knowledge to these new environmental problems.

Chapter 17

"Please Control Your Soul's Desire for Freedom"

Nick sat at his desk gazing at the water in the Niagara River as it raced under the Peace Bridge and past his office window. He was struck by the seemingly unending, perpetual rapid migration of such a massive amount of fresh water. He started to wonder if one of these water molecules passed this way once before during the time before the War of 1812 when this was the location of the Black Rock Ferry used by people migrating back and forth between Canada and the United States.

A group of his ancestors had crossed at this very spot in the early 1800s before the War of 1812. As his gaze wandered to the Peace Bridge, the present-day pathway for that same migration, he realized it was shut down due to the covid pandemic much like the War of 1812 shut down the ferry at this same location more than 200 years before. Not letting covid sour his thoughts, the nerd part of his mind returned him to more pleasant thoughts as the science was playing out before him.

Oh, he had spent endless hours pondering the pandemic and the butchering of the science by politicians and their paid, corrupted scientists. It hurt Nick's professional soul to realize what these evil people were doing to his profession. His public health background and 40-plus years of experience prevented him from turning a completely blind eye to the plain stupidity or planned corruption around him. He had spent a lifetime trying to keep his self-determination, especially in his later years when progressive socialism came out in the open.

Nick had identified it much earlier than most. He had spent much of his career working in, around, for, and at odds with government. He became aware especially in the 2000s how progressives seemed to be caught in a perpetual Do-LOOP of phony statistics backing up non-facts or non-facts backing up phony statistics, whichever the case. They may have been intellectually challenged, easily manipulated by "Madison Avenue" tactics, or just plain corrupt, but it seemed these people conflated, confused, contorted, or just made-up stuff and declared them as facts. He despised the phrase "follow the science," because he often was enraged knowing none of it was real science—some not even close.

"Please control your soul's desire for freedom" was the plea of the Chinese government, which issued this command to its population when—despite living in a totalitarian place—it started to openly rebel against total isolation during surges in the pandemic. Nick was concerned for his own country based on edicts put forth by federal and some state governments that clearly were not following science and were maddingly close to the actions of the Chinese. Can you imagine the feelings of scientists like Nick while watching the nonscience put forth by government "scientists" that were part of agencies that as a youth he once idolized? It was exasperating to say the least, especially when the media force fed this junk to an unsuspecting public made pliable by hideous scare tactics.

But all that did not enter his mind at this moment. Without consciously thinking about it, the theory of fluid dynamics and the study of how fluids behave when they're in motion was rolling around his brain. He was awestruck at how fast the water was moving as it left the large expanse of Lake Erie and entered the much narrower Niagara River. "Boy, that water is moving fast," he thought as it sped toward its destiny with Niagara Falls and eventually Lake Ontario. His favorite time to view this was when the ice boon that was placed in Lake Erie in the winter to prevent ice sheets moving down toward the power intakes was removed.

The ensuing result was a parade of miniature ice bergs floating along the river over the course of a few spring days. He wasn't quite sure—his knowledge of physics theory was rusty—but he figured the increased flow was due to two main reasons. He had read in a US Army Corps of Engineers document that as the river leaves Lake Erie it immediately narrows to a width of 1,500 feet and a depth of 17 feet as it reaches a rock ledge, which naturally controls its outflow. He knew the general bedrock direction and figured that gravitational potential energy was being transformed to kinetic energy. But mostly and certainly more obvious to him was that the larger expanse of the lake created a big pressure difference once it entered the narrow Niagara River, and the resulting net force produced this rapid acceleration of the water—or something to that effect. A fleeting memory of Bernoulli's equation rolled past his thoughts before his mind pushed on from science to history.

He began to think again about the migration of people in this area in historical times. Migration to Canada before the War of 1812 occurred from the population centers along the U.S. east coast. Nick had read in his family records a firsthand account of how some of his ancestors traveled in a covered wagon from the town of Washington, Duchess County, New York, to Canada. They traveled on a network of Indian trails and new roads that led through deep forests and along streams, including the Seneca trail, the Mohawk/Iroquois Trail, and the Great Genesee Road. The trip took them 21 days on the road.

In Cayuga, they crossed a wooden bridge over a mile long. At the Genesee River (Rochester today), the country was quite new with very few settlers. They traveled past Indian villages and crossed into Canada just north of the Buffalo Creek Settlement (Buffalo) at Black Rock.

Around the time Nick's ancestors traveled through Buffalo, there were a half dozen houses and about 25 people living in New Amsterdam, which eventually became Buffalo. By 1804, the caustic then-president of Yale College, Timothy Dwight, who published Travels in New England and New York, visited the village and is reported to have counted "about twenty indifferent houses." That same year, Erastus Granger arrived from New England and recorded that the village had several homes, a store, and a tavern. Granger settled in a log house on North Main Street and set out to complete the special mission with which he had been charged by President Thomas Jefferson; to establish Buffalo's first post office. Once Joseph Ellicott completed the "great survey," however, Buffalo's population grew rapidly. By 1810, the population was around 1,500, and a decade later it had risen to over 2,000.

Nick's ancestors crossed in a scow of about ten tons burthen, about 32 feet long by 8 feet wide, with four oars or sweeps and two men at each oar and one man with a longer sweep at the stern to steer. These ancestors and other families from Duchess founded one of the earliest and most successful Quaker communities in Upper Canada. Nick thought, "Here I am, sitting in my office, possibly at the very spot my ancestors traveled before they crossed into Canada."

As Nick sat in his office at the foot of Busti Avenue looking out at the Peace Bridge and the rapid migration of water from the lake, he contemplated the name of the street. It was named after Paolo Busti, a native of Milan, Italy. Busti reportedly came to America from Amsterdam around 1798, and because of his role as the general agent of the Holland Land Company, many considered Busti—rather than Joseph Ellicott—the founder of Buffalo. Apparently, Busti directed all the work of Ellicott, the chief surveyor of the Holland Land Company's purchase in New York State.

Not a native of Buffalo, Nick began thinking about his many years in Buffalo. When he left his hometown, his friends would ask, "Why Buffalo of all places?" He came to love Buffalo, and he sat there realizing that he had lived in the Buffalo area longer than any other place, including his childhood and young adult home outside of Peekskill, New York. He still was drawn to the beauty of the Hudson Highlands, especially how the air breathed easy when he was in the hills of his hometown in the cool evenings.

He had recently become more interested in the history of Buffalo and learned that in 1664, King Charles II gave the territory, which included Buffalo, to James, Duke of York. It was marked on a map previously created by the explorer Baron Lahontan as "Fort Suppose." The map indicated that the area was inhabited by thousands of American Indians of the Neuter, Erie, and Seneca nations.

Lahontan was born to a noble French family close to the Town of Pau near the Pyrenes. Around the age of 17, Lahontan came to French Canada and served as a soldier and translator, and he traveled extensively. He recorded some of the earliest written accounts of the other native tribes—Ottawa, Huron, Iroquois, Illinois, and many more—that populated New France. Some say that a book based on his work, New Voyages to North America (London, 1703), is an extraordinary account of Native Americans at that time with unusual details, spectacular engraved illustrations, and a unique description of these people.

Nick learned that many of the first buildings in Buffalo, including the first brick structure—the home of William Hodge at 1358 Main Street—were built in what became known as the Cold Spring District on the outskirts of the main city area nearer to the lake. It is still known as the Cold Spring District today on City maps, but few know the history of its name. The name Cold Springs originated from the sparkling natural spring that furnished pure water for inhabitants from miles around.

William Hodge described Cold Spring as "a large basin, surrounded by bluff banks excepting on the north-east side, where the pure, cool stream flowed forth. ... Originally the banks were somewhat sloping, and steps were cut into the steepest part ... by which to go down to a plank which extended several feet over the bubbling and boiling water. Lying stretched out on [this] plank, face downward, many a one has slaked his thirst in days gone by, and from that plank, many a pail and jug has been filled with the pure, cool beverage." Cold Springs was the very place that Nick's ancestors described in their journal as the place they stopped on their way to Canada to fill their water barrels.

Nick recalled how when he worked at E&E his roommate Mike and his friend Carmine did an extensive study on a log road discovered under Main Street in Buffalo. In June 1980, during initial construction of Buffalo's Light Rail Rapid Transit System, parallel logs were discovered below the present Main Street surface in downtown Buffalo. Preliminary investigations suggested that they were part of a log or plank road used sometime during the first part of the nineteenth century. History suggests that the first road into Buffalo was an Indian trail that later became known as the "Great Central Trail" and was located along the approximate route of what is now Main Street. This trail crossed New York State from the Hudson River to Lake Erie.

During colonial times, the trail was used to bring supplies to the Holland Land Company's field crews. In 1801, Joseph Ellicott commissioned a new trail along the same line as the "Great Central Trail," which became known as "Buffalo Road" and later Main Street. This road was an ungraded, stump-covered wagon route from what is now Batavia, New York, to the Village of New Amsterdam (now Buffalo). Historical reports suggest that the road from Three-mile Creek to the "Cold Springs" was made of a log causeway.

Jedidiah found the information on the log road detailed in technical reports, and a summary was presented in 1984 in volume 13 of the journal Northeast Historical Archaeology.

Nick recalled reading that during that time, in the early 1800s, Syracuse, New York, was the nation's salt production capital, and salt was like gold. The firm Porter, Barton & Company controlled transportation on the Niagara River. Starting around 1805, they used Durham boats—large wooden, flat-bottomed, double-ended freight boats—to carry salt to Black Rock (now part of Buffalo). Black Rock was known as the great salt exchange and was the commercial center for salt merchants. The salt was eventually transported from Black Rock to Erie,

Pennsylvania, and eventually Pittsburg and other areas across America. He recalled reading how during the War of 1812 British ships would try to capture these salt boats as they scurried up the nearby rivers and streams for safety from the bigger British ships.

The name Black Rock came from the black rock in the Niagara River on a plateau where the Black Rock Ferry was situated—just down from where the Peace Bridge was in 2022, when Nick sat in his office overwhelmed by the pace and energy of the river below. Back in the 1800s, the "Rock"—as it was called—projected into the river and was removed when the Erie Canal was built. When the wind blew down the lake, the vessels running from the Black Rock Ferry were wind-bound at "The Rock."

It was not uncommon for five thousand to six thousand barrels of salt to pile on the piers and shore waiting for the weather to change. At that time, "The Rock" was covered with salt merchants from Pittsburg, captains, and boatmen who would meet at the tavern. The old tavern at "The Rock" was as distinguished along the frontier in those times as Wall Street and other business centers are today. As Nick sat in his office in the early morning on a cold and windy mid-December day in 2021, he thought back on the past couple years and the many years that preceded them. It was just about to tick into a new year, and the world was still reeling from that past two-plus years of the covid pandemic, which was being accentuated by a new virus variant more communicable but less virulent.

Because of his public health training and many years of experience, he knew other variants would follow. He would be 67 in a few months, and the small firm he was part owner of had "weathered the covid storm" through the hard work of him and his partner and their staff. Despite the times, the two partners were still rugged individuals and smart enough to have worked through the whole mess—from the disease's spread out of China to the exasperating response by the leaders in the United States. Both highly educated and experienced, their overriding traits were heightened by their intellectual integrity and work ethic. They were preparing to open a branch office in Florida.

Florida was alluring for many reasons, but Nick chose not to dwell on negatives today as his mind wanted to play in history and science. Aristotle said, "For the man who flies from and fears everything and does not stand his ground against anything becomes a coward, and the man who fears nothing at all but goes to meet every danger becomes rash; and similarly the man who indulges in every pleasure and abstains from none becomes self-indulgent, while the man who shuns every pleasure, as boors do, becomes in a way insensible; temperance and courage, then, are destroyed by excess and defect, and preserved by the mean." Finding the mean also means risking failure but minimizing the risk. The two partners were risk takers, otherwise they never would have started a business. Those moments that scare us, require a bit of bravery, and might even result in taking a tumble from the risks. Growth through difficult circumstances requires risk. To quote Clay Hayes, a contestant on season 8 of the show Alone; "The world would be a better place if once in their life people got so hungry that they had to suck the eyeballs out of a rabbit's skull." Maybe that's a bit extreme, but the loss of rugged individualism was evident in America in the 2000s.

Nick thought about his career and how society jumps from one issue to another, seemingly leaving the previous one vanquished or minimally conquered. The same was true for environmental concerns as the headline society pinballed from one or two crises to the next, leaving the pieces behind for the serious people to contend with. Nick thought back through the various new environmental concerns making headlines in every era. DDT and the eggshells of eagles, dioxin, urea formaldehyde foam insulation, PCBs, asbestos, radon, lead-based paint, mold, and now PFAS. They were all real issues despite the hyperventilation that comes and goes and comes back again as some new generation discovers its industrial past and present.

Like so many, he too reveled in the amazing products that resist heat, oil, stains, grease, and water, which dazzled the consumers of his day. Water-repellent clothing, furniture, adhesives, paint and varnish, food packaging, heat-resistant nonstick cooking surfaces, and insulation of electrical wires are just some of its products and uses. Even though their unique and useful properties had been in use in industry and consumer products since the 1940s, the rapid interest in the identification of the next "chemicals of emerging concern" resulted in the need for experts to identify them and reduce exposure to humans and the environment.

Now in the twilight of his career, Nick began to reflect on the many different projects and sites he worked on. He had worked on some of the more consequential sites in the early days of the industry, from Love Canal to the Lipari Landfill and many others. At each site, he had some exposure and there was no way to know if any or the cumulative effects would bring him down. Later in his career, he would spend hours on petroleum- and chlorinated solvent–contaminated sites completing forensic environmental assessments, which he excelled at.

He told a young colleague once using the story as a teaching moment that with all the nasty and dangerous Superfund hazardous waste sites he worked on, he may have gotten his worst exposure just a few years before on a brownfield site he was remediating. At this site, the crew had uncovered a pipe that had led from a previously removed xylene tank, and Nick had gone over to test the air from the broken end. Despite wearing a mask with organic vapor cartridges, the levels were so high that the chemical quickly overwhelmed the cartridges. Nick cursed himself as he backed away quickly upwind. The organic vapor monitor he was carrying was shrieking its steady alarm warning that the levels were excessive. At the time, Nick knew he had received a significant dose, and he was mad at himself for the underestimation and his mistake. Later it was a perfect teaching moment for Nick to share with the younger staff.

He sat at his desk, focusing on the past few years. Covid was killing people. He wondered if the cumulative exposures he got over the past forty-plus years of his working life would result in his eventual and certain death. Did he get a one-time exposure, say on Johnston Atoll, that sealed his fate all these years later, or was it the dose he got on the more recent Brownfield site? He had spent close to a month on that island, where biological and chemical weapons had been stored and leaked and radiological material had been spread from the nuclear weapons testing—not to mention his job of sampling the coral drenched in Agent Orange he sampled for dioxin especially in and around the burn pit.

He had just come across an article on military exposure, and he read something about Johnston Atoll being included in the congressional bill H.R.812—"Agent Orange Equity Act of 2011" during the 2011–2012 Congressional year. This Act included veterans who: (1) served on Johnston Island during the period beginning on April 1, 1972 and ending on September 30, 1977; or (2) received the Vietnam Service Medal or the Vietnam Campaign Medal. The Act was associated with "Exposure to dioxin, or other toxic substance in a herbicide or defoliant, during the conduct of, or as a result of, the testing, transporting, or spraying of herbicides for military purposes."

Although Nick and his crew were neck deep in crushed coral on JI that was laced with dioxin from herbicide orange, they were non-military contractors and not "active-duty military," so he was sure this probably did not apply to him and his crew all those many years ago. His thoughts drifted back to his friend John

Duwaldt, who he had just found out died of some form of cancer in 2019 and was with him in the trenches at JI. Then his thoughts drifted to his other friend James B. Moore "Himself," who had served there on active duty and was also gone. James B. Moore "Himself" regaled Nick and others with the "war" stories of serving on Johnston Island.

Perhaps it was the eight years he spent on the Department of Energy (DOE) radionuclear facility that would get him. The West Valley project was a demonstration project to take the waste from nuclear waste repossessing and turn it into glass logs using a vitrification process. Nick thought, "In the end, no matter what the reason or culmination of reasons—exposure, heredity, socio-economic status, diet, habits, or whatever—you can't get out of this world alive. Time is short, and the history of my work is the story of my life. That is the story of everyone's life." Then he smirked as he recalled what the venerable and profound 49th vice president of the United States recently said: "It is time for us to do what we have been doing. And that time is every day."

A band of geniuses? That's how someone had recently described the early days at E&E, where he and others had come together at the beginnings of their careers. Although some were certainly geniuses, Nick only gave himself credit as being one of just a collection of smart and widely diverse individuals. Environmentalist? No, certainly not. Although some may have been, most were environmental professionals and capitalists whose mission was to protect the environment and workers while industry advanced the nation. Tree huggers? No. Although they were experts in bugs and bunnies, civil and environmental engineering, planning, chemistry, atmospheric science, biology, microbiology, archaeology, physics, and mathematics, they certainly were not the environmentalists of today—at least most weren't.

Nick reflected more about the "wars" he went to with his many colleagues on all those Superfund sites years ago. For sure what they did and their experiences although uniquely dangerous were no comparison to being a participant or even a bystander in an actual fighting war. Living through the Vietnam era as a young man—having it stream into his home nightly on TV and watching his brothers' friends and others he knew come home very changed—made him understand that. He had his own form of physical combat on the playing fields and gym mats, and certainly that was not even a close comparison. He always had true respect for the men and women who fought in war, whatever their reasons.

But still, it's hard to describe the feeling of walking into the chemical unknown on each unique hazardous waste site, usually stuck in some awful corner of a city or in the "bad" parts of towns or isolated almost without warning in some far rural place, farmers field, at the town landfill, or some other uncategorizable place. The first visit at each site was unique, and some of his experiences created flashbacks because of the intensity of them; walking among rows of leaking drums stacked two to three or four high, some "jumping" a few inches with a bang as the heat of the day expanded their contents. Walking across multicolor sites with red, green, orange, blue, and brown oozing under each boot step or along the stormwater ditch that channeled the flow or in some dark and dank industrial building long ago abandoned as manufacturing was allowed to walk out of America for the new "service" industry.

For a few years after his return from Johnston Island, he would flash back unexpectedly—maybe as he watched some special on islands or when he was in a particularly hazardous environment. Despite the fun things they did to keep their minds occupied while on the island, there was always the ever-present military mask and vials of atropine they were required to have within arms' reach made worse by their daily trek down to the most downwind place on the island and the burn pit where the agent orange was partially destroyed. He had not realized until years later and after a few vivid flashbacks how much that island time had affected him.

Although the actual tasks and the early hazardous waste jobs varied, they all revolved around assessing, investigating, and sampling air, water, soil, or actual waste and then designing and then remediating the "bad" stuff they found. The mission was always to understand the risk to the environment and humans or to stop and clean up a spill in progress. Nick and others in his industry were the first responders. Many times they were also the last responders because their knowledge and that gained from their work would hopefully result in the ultimate remediation. In the beginning it was all new; no one had done it before. They were brand new programs. Whether they walked into long ago abandoned industrial buildings and flooded pitch-black cold basements or into a farm field surrounded by drums, they were always surreal places for sure.

In the early days, just after he graduated from Yale, he held the scientists from the Center for Disease Control (CDC), the National Institutes of Health (NIH) and the World Health Organization (WHO) with the highest regard. Now, in the waning time of his career after spending many years working around government agencies and departments, he was somewhat jaded to say the least. The past few years of the blatantly obscene reaction to covid by overpaid, incompetent, and possibly corrupt government "scientists" angered Nick to an extreme—like so many other things politicians touched. It really hurt his soul to realize that these venerable agencies and scientists once thought of by Nick as representing scientific excellence had been turned into corrupt entities by political scientists and America's political leaders. The politicization of public health meant that the credibility painstakingly built by the CDC, FDA, and WHO in Nick's youth was irrevocably and irrefutably broken.

Nick thought again about a recent health episode he had that occurred unexpectedly. He had wondered if he had "woke up dead" and was now living in an alternate universe. He first heard the expression "woke up dead" from his childhood friend and high school teammate Mike Twardy. They went to college together to wrestle. Mike told him a story while they sat and drank some beers in Mike's room at Cashin Hall. This was the dorm they both lived in during their freshman year; the place where they witnessed some of the crazier events of college life at a large university during the early 1970s as real-life events similar to those described in Robert H. Rimmer's novel The Harrad Experiment (1966) played out. Nick and Mike were greeted at UMASS with coed dorm suites, coed bathrooms, and an openly promiscuous environment. When they were at UMASS, the school had held the streaking record for one whole weekend as hundreds of students took a nude run (and walk) across the campus starting and ending at the student union before some other large college broke it a few days later. Nick recalled how on any given evening he would look out and see some streakers followed by some more trying to outdo the last, including one guy who rode a unicycle nude around the complex. Maybe it was just the end of the era of the out-of-control 1960s or just a way for young people to shed the darkness of the Vietnam era.

Cashin was one of the newer dorms on campus at that time. It was one of the three high-rise towers in a secluded complex composed of 8 or 9 floors each in what was referred to as the Sylvan Residential Area, which offered suite-style living in a shady wooded area. Each residence hall contained 64 suites. The other two dorms were Brown and McNamara Halls. The other undergraduate dormitories on this large university were Central, where they would move the next year, North Apartments, Northeast Area, Orchard Hill, and Southwest. The Southwest Residential Area, the largest on campus, was almost at the opposite side of the campus from his Sylvan complex. It housed almost 6,000 students in its five high-rise towers and eleven low-rise residence halls.

It was the first thing you'd see as you entered UMASS from the main drag. When he first saw this massive dorm complex, he thought of Co-op City in New York City. The look and feel were almost the same from afar as what Nick often saw when driving a particular route into New York City.

Mike was telling Nick that some kid he heard about had too much to drink and suffocated on his own vomit and "woke up dead," as only Mike would say in his artful way of expressing things. Now, as Nick sat in his office all these many years later thinking back on his life and long career, he wondered about "waking up dead."

· · ·

I always consider it positive to stand up and speak out against injustice—especially the kind that is ostensibly cloaked in justice. We have all been silent way too long. We can all hide under a rock, or we can stand up against the progressive nonsense that is fundamentally transforming a once great country. The first step is recognizing it, acknowledging it, and speaking out against it. You can run away and decide people are complaining when they elevate the injustice. Today, for example, I see the same racism and bigotry conducted by the same people I saw in the 1960s and on films in the 30s–40s. The only difference is they rearranged the deck chairs—but it's still the same racism and bigotry only with different sounding names that have been conflated and contorted by the same people.

My name is Jedidiah—a biblical name that traces back to an ancestor long ago. I was finishing up this detailed review of Nick McCarthy for his ancestor Bailey. I am a master historian and familiar with the geographic and historical record of places with extensive knowledge of genealogy and DNA database research. I had spent a lot of time on the history but equally as much on the DNA science and hereditary history of Nick and his ancestors.

In 1985, DNA entered the courtroom for the first time as evidence in a trial, but it wasn't until 1988 that DNA evidence actually sent someone to jail. As an expert in this field, I know DNA is a complex area of forensic science that relies heavily on statistical predictions and has only gotten much more sophisticated with time. In early uses of DNA in criminal trials, attorneys often overwhelmed jurors with heavily laden scientific terms and mathematical formulas that only a few understood. These were easily undermined by crafty defense attorneys. Many advances in DNA research have occurred since then, including new testing procedures and more scientifically defensible methods of extracting DNA. The biggest advancement, however, has been the sheer volume of data and expanded DNA databases that allows for much better statistical application and inferences with manageable standard deviations.

DNA profiling is now a cornerstone tool in genealogy. Once the use of Y-SRT testing to determine paternal lineage and mtDNA (mitochondrial DNA testing) to determine maternal lineage became more advanced, DNA sequencing became the single greatest scientific breakthrough in bioinformatics and genetics, especially testing that focuses on different DNA markers.

I had just finished researching a seemingly less hazardous event that happened during Nick's later stages of his career. This event was somewhat nondescript, as were many of the work tasks Nick had over his 40-plus year career. He was meeting one of his clients who was a relatively well-known local developer who took old vacant buildings and retrofitted them into new usable spaces—usually some form of mixed residential, hotel, and commercial space. He stood on Washington Street watching his client rush across the street to meet him. His client had called him just 45 minutes before, asking if he could do a quick Phase I Environmental Site Assessment because the bank needed it for a closing in two weeks.

This was an old 4- or 5-story former retail warehouse that had been vacant for a number of years smack down in the center of the city. He had done several jobs like this for this client and similar jobs for other clients. When he was not working on cleaning up an old gas station site or dry cleaners, he was walking around in some old, abandoned building or house. The basements were always his least favorite for many reasons.

As the client approached, he unlocked the door and said to Nick, "Call me when you are done, and I'll lock up." Typically, Nick preferred to have the client, or a client representative, walk him around. Although he had done many by himself, he usually never did a large, old vacant building like this by himself. It would not be the first time when he walked in on squatters in the pitch dark or barely missed walking into a submerged elevator shaft in a dark basement with a foot of water or found needles and crack paraphernalia, rats, dead pigeons—and live ones flying directly at his head—and more.

He swore a couple of times he had gotten a case of psittacosis from breathing in bacteria from pigeon poop that was layered an inch thick in the stairwells. He only found out later that he had developed asthma probably from an allergy to these things as he walked over asbestos-laden debris, lead-based paint chips, PCBs, mold, dust, and petroleum from the engines that ran the boilers.

Nick recalled the one time he had a job to assess about forty vacant homes on the east side of Buffalo on about three or four scattered streets. He had just been watching a couple of those "ghost investigation" shows for some reason and now regretted it as he went into one spooky vacant boarded-up house after another. In one particular house, he was about to enter a dark attic that someone had converted into a room. As he opened a makeshift door and shinned his flashlight, he saw a face right in front of his eyes. Well, he almost screamed like a little girl until he collected himself and realized what it was. Someone, maybe in the forties or fifties, had pasted magazine covers and pin-up girls across the low ceiling, and one had slightly peeled off and was hanging straight down. It was the face that Nick was peering at. He ran into a few vagrants on this assignment, and after one he came home and gave himself a lice bath.

The occurrence, development, and events of Nick's life happened seemingly by chance—as most lives do—or perhaps it was just ironic cruelty. How else would you describe the outcome of life mirroring academic study? The friends and acquaintances he met living his life's purpose foretold their collective outcome. "You can't get out of this world alive," which was never a consideration during Nick's and his friends' youths. Was the outcome of their lives brought on by the things they later studied? Did these put them in the position for the pathways that most likely ended their lives, or was there much more to it?

Nick went to University of Massachusetts despite the football coaches at Hofstra and University of Albany indicating they wanted him to come to those universities. Instead, because he liked the campus so much more, he went with his friend and high school teammate Mike Twardy to wrestle, believing he would just walk on the football team.

He had received a rejection notice from UMass stating that the state university system only allowed 10% out-of-state students. That all changed with a call from the wrestling coach telling him he got him in but had to change his major to liberal arts. The coach followed that by saying that if he couldn't get Nick in with his grades, he would never be able to get the rest of his recruits in.

At school, he and Mike would take long trips on Mike's motorcycle up into the Berkshire Mountains on old logging roads deep in the woods. In the summers he worked on the garbage truck with his best friend Scott and hung with his friends long before he ever dreamed of the future career he would have. Now, at the end of his career, he was working with a new generation of experts, their genius advancing new technology in environmental protection and remediation, fostering the advancement of civilization. You never stop learning, which was something Nick adored about life and what made him work much longer than others. "Why should I retire?" he would reply when asked by friends. "What else would I do? I love this stuff!"

As Jedidiah sat back reflecting on all the information he had accumulated for his detailed report on Bailey McCarthy's ancestor Nick, he poured himself a Michter's and sat back in his leather arm chair content with his work. He thought about his eminent move back to the Cortlandt area and his beloved Hudson River home to rekindle his life there. But he knew he would never forget Buffalo, a place he had come to love and respect. As he sat back to enjoy his bourbon, he gazed out at the perpetual flow of water from Lake Erie into the Niagara River and past his window view. He had rented a top-floor suite in the building that once housed Nick's office.

As he watched the rapid, unending migration of water, he reached over and found the specific Stan Getz vinyl album he had just purchased and slipped it onto his prized turntable. He was amazed when earlier that day he found this pristine original just by chance in that old record store on Jefferson Avenue near Best Street. That area of Buffalo, the east side, was really so different than back in Nick's time when, at the end of his career, he had a number of projects because money was being pumped in to revitalize this long-forgotten part of town.
One of many obscure but enlightening facts that Jedidiah learned about Nick was that one of his favorite songs was a Stan Getz favorite. The record set he bought earlier was the five-LP box set that contained nearly all of Stan Getz's classic bossa nova sessions, five wonderful yet diverse LPs including Jazz Samba, Big Band Bossa Nova, Jazz Samba Encore, Stan Getz/Laurindo Almedia and Getz/Gilberto.

He sat back in the chair again, sipping his drink when the cool-toned music of Stan Getz began with composer/pianist Antonio Carlos Jobim and guitarist João

Gilberto singing four verses of lyrics in Portuguese. As his gaze focused back on the water, the warm feeling of the bourbon easing his edge even more, the sensual rhythmic sound of singer Astrud Gilberto began "The Girl From Ipanema," and he thought of Nick McCarthy's life.

> Tall and tan and young and lovely
> The girl from Ipanema goes walking
> And when she passes, each one she passes goes, ah
>
> When she walks, she's like a samba
> That swings so cool and sways so gently
> That when she passes, each one she passes goes, ah
>
> Oh, but he watches so sadly
> How can he tell her he loves her
> Yes, he would give his heart gladly
> But each day, when she walks to the sea
> She looks straight ahead, not at him

"The water keeps migrating," thought Jedidiah. I wonder if one of the molecules passing right now passed here when Nick was around?" The Niagara River and Falls have been known to Europeans outside of North America since the late 17th century, when Father Louis Hennepin, a French explorer, first witnessed them. He wrote about his travels in "A New Discovery of a Vast Country in America" (1698). He then recalled the exploits of French explorer Rene-Robert Cavalier, Sieur de La Salle, who launched the first sailing ship, the Griffon, built in 1679, to ply the upper Great Lakes. La Salle was seeking the Northwest Passage to China and Japan.

Can you imagine what these were to the "natives" and what they thought when they first witnessed them? Or perhaps the first Europeans to witness them were many centuries before by the Vikings. Maybe the Knights Templar came to this spot a hundred years before Columbus "found" America because they "would not Control Their Souls' Desire for Freedom."

Appendix – Additional Technical Detail

JOHNSTON ISLAND

Johnston Atoll is one of the most isolated atolls on the planet and is the result of 70 million years of volcanic eruptions, limestone capping, and reef growth. The four small islands of Johnston Atoll are home to over 200 species of fish, 32 species of coral, and 20 species of native and migratory birds. The climate is flawless in terms of offering consistent, hot, balmy Pacific summer days.

Today there is no longer a 14,000-foot runway, barracks, dining hall, PX or Officers Club, laundry, volleyball court, swimming pool, golf course, or softball field. If you're thinking of visiting the atoll, you may find it tough: Public entry to the islands is by special use permit only issued by the U.S. Airforce and generally restricted to scientists and educationists.

Between 1946 and 1958, the US military conducted 67 nuclear tests at Bikini and Enewetak Atolls in the Marshall Islands. Less well known are the US nuclear tests on Johnston Atoll in 1962. A timeline of some of the more consequential events associated with the military mission on the atoll that affected environmental concerns on the island is summarized below. It was only many years later that Nick found out about these various operations and the potential exposure he may have gotten during his time on island.

The original military buildup on the atoll included the construction of a 4,000-foot by 500-foot runway/airfield on Johnston Atoll as well as two 400-man barracks, two mess halls, a cold-storage building, an underground hospital, a fresh-water plant, shop buildings, and a fuel storage area. On the day the Japanese struck Pearl Harbor, December 7, 1941, the USS Indianapolis was out of her home port of Pearl Harbor to make a simulated bombardment at Johnston Atoll.

Japan's strike at Pearl Harbor occurred as the ship was unloading marines, civilians, and stores on the atoll. On December 15, 1941, the atoll was shelled outside the reef by a Japanese submarine, which had been part of the attack on Pearl Harbor eight days earlier. Several buildings, including the power station, were hit, but no personnel were injured. Additional Japanese shelling occurred on December 22 and 23, 1941. On all occasions, Johnston Atoll's coastal artillery guns returned fire, driving off the sub. By 1943, the runway was lengthened to 6,000 feet, and the island was enlarged. There were many times when the runway was needed for emergency landings for both civil and military aircraft, including one landing by a Qantas Boeing 747.

During WWII, Johnston Atoll was used as a refueling base for submarines and also as an aircraft refueling stop for American bombers transiting the Pacific Ocean, including the Boeing B-29 Enola Gay. During that time, the atoll was one of the busiest runways in the Pacific with air transports stopping at Johnston en route to Hawaii.

Between 1958 and 1975, the Johnston Atoll area was used as an American national nuclear test site for atmospheric and high-altitude nuclear explosions, including the "Hardtack I" nuclear test. Two nuclear tests occurred during 1958 codenamed "Teak" and "Orange," involving 3.8-megaton explosions from rockets launched from Johnston Atoll. The first operational ballistic missiles deployed by the U.S. Air Force were launched from Johnston Atoll in 1962 as part of "Operation Fishbowl" under the "Operation Dominic" Pacific Ocean nuclear tests. The "Fishbowl" series included four failures, all of which were deliberately disrupted when the missiles systems failed during launch and were aborted.

Another Fishbowl series test, carrying an active warhead, had to be destroyed 10 minutes after liftoff. Subsequent nuclear weapon launch failures from Johnston Atoll caused serious contamination to the island and surrounding areas, scattering radioactive debris over the island contaminating it, the lagoon, and Sand Island with weapons-grade plutonium and americium that remains an issue to this day.

From 1963 to 1970, the Navy's Joint Task force 8 and the Atomic Energy Commission (AEC) held joint operational control of the island during high altitude nuclear testing operations. Also, in 1965, the island was associated with biological warfare testing involving a new virus discovered in birds by teams from the Smithsonian's Division of Birds in the Pacific. First isolated in 1964, the tick-borne virus was discovered in ticks in a nest of common noddy terns at Sand Island, Johnston Atoll. It was designated Johnston Atoll Virus and is related to influenza viruses. By April 1965, Johnston Atoll was used to launch biological attacks against U.S. Army and Navy vessels 100 miles Southwest of Johnston Atoll. The biological agents released during this test were the causative agent of Tularemia, the causative agent of Q fever, and Bacillus globigii (Agent BG). Ships equipped with spray tanks released live pathogenic agents in nine aerial and four surface trials.

In 1970, Congress changed the island's military mission as the storage and destruction of chemical weapons. Johnston Atoll became a chemical weapons storage site in 1971, holding about 6.6 percent of the U.S. military chemical weapon arsenal with the re-deployment of the 267th Chemical Company and consisted of rockets, mines, artillery projectiles, and bulk one-ton containers filled with Sarin, Agent VX, vomiting agent, and blister agent, such as mustard gas.

The chemical agents were stored in the high security Red Hat Storage Area (RHSA) along the southwest portion of the island in hardened igloos (bunkers). Agent Orange was brought to Johnston Atoll from South Vietnam and Gulfport, Mississippi, in 1972 under Operation Pacer IVY and stored on the northwest corner of the island known as the Herbicide Orange Storage site, which was dubbed the "Agent Orange Yard." The Agent Orange was eventually destroyed on the Dutch incineration ship MT Vulcanus in the Summer of 1977. The Environmental Protection Agency (EPA) reported that 1,800,000 gallons of Herbicide Orange were stored at Johnston Atoll in the Pacific and that an additional 480,000 gallons stored at Gulfport, Mississippi, was brought to Johnston Atoll for destruction. Leaking barrels during the storage and spills during re-drumming operations contaminated both the storage area and the lagoon with herbicide residue and its toxic contaminant 2,3,7,8-Tetrachlorodibenzodioxin. This is the area Nick worked on when he arrived in 1984.

After Nick left the island, base closure operations began. The first step was developing a system to destroy all the chemical weapons that remained on the island. The Army's Johnston Atoll Chemical Agent Disposal System (JACADS) was the first full-scale chemical weapons disposal facility. It was built to incinerate chemical munitions on the island. Planning started in 1981, and construction began in 1985. All of the chemical weapons stored on Johnston Atoll were incinerated at JACADS by 2000, followed by the destruction of legacy hazardous waste material associated with chemical weapon storage and cleanup. JACADS was demolished by 2003, and the island was stripped of its remaining infrastructure and environmentally remediated. A monument dedicated to the only JACADS employee who died as a result of exposure to the nerve agent, Harry Sacks, was erected at the site.

In 2003, structures and facilities, including those used in JACADS, were removed, and the runway was marked closed. The last flight out for official personnel was June 15, 2004. After this date, the base was completely deserted. The only structures left standing were the Joint Operations Center (JOC) building at the east end of the runway, chemical bunkers in the weapon storage area, and at least one Quonset hut. Rows of bunkers in the Red Hat Storage Area remain intact; however, an agreement was established between the U.S. Army and EPA Region IX on August 21, 2003, that the Munitions Demilitarization Building (MDB) at JACADS would be demolished and the bunkers in the RHSA used for disposal of construction rubble and debris. After placement of the debris inside the bunkers, they were secured and the entries blocked with a concrete block barrier (a.k.a. King Tut Block) to prevent access to the bunker interior.

REFERENCES

1) A Letter to the National Academy of Sciences
https://www.nas.org/academic-questions/35/3/a-letter-to-the-national-academy-of-sciences

2) Superfund Record of Decision – Whitehouse Waste Oil Pits (Amendment) Fl, 1992
https://nepis.epa.gov/Exe/ZyPDF.cgi/91002WHN.PDF?Dockey=91002WHN.PDF

3) Public Health Assessment for Whitehouse Waste Oil Pits Whitehouse, Duval County, Florida Cerclis No. Fld980602767 September 14, 1992. U.S. Department of Health and Human Services
https://www.floridahealth.gov/environmental-health/hazardous-waste sites/_documents/w/whitehousewasteoilpit091492.pdf

4) *Behind enemy lines in Vietnam,* by DON MOORE
Behind enemy lines in Vietnam | War Tales (donmooreswartales.com)

5) Superfund Sites in Reuse in New York; USEPA
Superfund Sites in Reuse in New York | US EPA

6) USEPA Superfund Site Love Canal, Niagara Falls, NY
LOVE CANAL | Superfund Site Profile | Superfund Site Information | US EPA

7) American Cyanide Site – 2011 Fact Sheet – New Jersey Department of Environmental Protection amcy (nj.gov)

8) NYSDEC - Per- and Polyfluoroalkyl Substances (PFAS)
https://www.dec.ny.gov/chemical/108831.html

9) Health Consultation-Petro Processors of Louisiana, Inc. Post-Hurricane Groundwater Sampling Evaluation East Baton Rouge Parish, Louisiana EPA Facility Id: Lad057482713 September 30, 2006. U.S. Department Of Health And Human Services Microsoft Word - Petro Processors jk cmts.doc (cdc.gov)

10) Federal Remediations Technology Roundtable - Incineration at the Petro Processors Superfund Site, Baton Rouge, Louisiana. Federal Remediation Technologies Roundtable Cost and Performance (frtr.gov)

11) REGION 6 CONGRESSIONAL DISTRICT 06 East Baton Rouge Parish Updated 8/9/02. Petro-Processors of Louisiana, Inc. Louisiana EPA ID# LAD057482713 NPL FACT SHEETS (INITIAL TITLE) (epa.gov)

12) Superfund Contracts – USEPA November 1990 91003GO8.PDF (epa.gov)

13) EPA History (1970-1985) EPA History (1970-1985) | About EPA | US EPA

14) Oil Spills and Spills of Hazardous Substances. USEPA 1975 2000TXLH.PDF (epa.gov)

15) Superfund 25th Anniversary Oral History Project - STAN KOVELL EPA Staff Member Who Initiated EPA's Contract Lab Program Date: October 5, 2005 Location: Arlington, Virginia Microsoft Word - KovellHC.doc | US EPA ARCHIVE DOCUMENT

16) Lipari Landfill Cleanup – New York Times Cleanup Is Set at Jersey Dump, the Nation's Worst - The New York Times (nytimes.com)

17) Lipari Landfill Superfund Site Remediation Pitman, New Jersey Date of Execution: March 1994. Sevenson Environmental Services Lipari Landfill Superfund Site Remediation - Sevenson Environmental Remediation (archive.is)
18) Superfund Site Lipari Landfill Mantua Township, NJ Cleanup Activities USEPA LIPARI LANDFILL | Superfund Site Profile | Superfund Site Information | US EPA

19) E&E - Decades-old N.J. cleanup a cautionary tale for Pruitt. BY: Corbin Hiar 09/05/2017 SUPERFUND: Decades-old N.J. cleanup a cautionary tale for Pruitt -- Tuesday, September 5, 2017 -- www.eenews.net

20) Superfund Record of Decision – Lipari Landfill, NJ (Second Remedial Action 9-30-85) USEPA September 1985 9100SAHA.PDF (epa.gov)

21) E. A. Dienemann, R. C. Ahlert & M. R. Greenberg (1991) REMEDIATION OF THE LIPARI LANDFILL, AMERICA'S #1 RANKED SUPERFUND SITE, Impact Assessment, 9:3, 13-30, DOI: 10.1080/07349165.1991.9725715 To link to this article: https://doi.org/10.1080/07349165.1991.9725715
REMEDIATION OF THE LIPARI LANDFILL, AMERICA'S #1 RANKED SUPERFUND SITE (tandfonline.com)

22) Toxic NJ: Lipari Landfill. Russ Zimmer Asbury Park Press. December 5, 2017
Toxic NJ: Lipari Landfill (app.com)

23) Johnston Atoll Base Guide 1984

24) Wikipedia - https://en.wikipedia.org/wiki/Johnston_Atoll

25) Weapons of Mass Destruction (WMD) - Johnston Atoll/Kalama Atoll
SOURCE: Global Security
https://www.globalsecurity.org/wmd/facility/johnston_atoll.htm

26) Wikipedia - Johnston Atoll Airport -Johnston Atoll Airport

27) Wikipedia - Shelling of Johnston and Palmyra- Shelling of Johnston and Palmyra

28) Wikipedia - Operation Fishbowl, Operation Dominic and Operation Hardtack
Operation Fishbowl, Operation Dominic and Operation Hardtack I

29) Project SHAD, Project 112, and Desert Test Center- Project SHAD, Project 112 and Deseret Test Center

30) Secret Bases – Johnston Atoll
http://thelivingmoon.com/45jack_files/03files/Johnston_Atoll_02.html

31) Johnston Island circa 1991 - Johnston Island circa 1991 - YouTube

32) "Cleaning up Johnston Atoll", APSNet Special Reports, November 25, 2005, https://nautilus.org/apsnet/cleaning-up-johnston-atoll/

33) Spy -The Spy Who Got Away: The Inside Story of Edward Lee Howard, the Man who Betrayed His Country's Secrets and Escaped to Moscow. David Wise
34) A new look at the .22-Caliber Killer | Local News | buffalonews.com A True Story of a Serial Killer, Race, and a City Divided Catherine Pelonero

35) Five-Year Review Report Second Five-Year Review Report for Envirochem Site Zionsville Boone County, Indiana April 2008 PREPARED BY: United States Environmental Protection Agency Region 5 Chicago, Illinois-
https://semspub.epa.gov/work/05/295754.pdf

36) Remedial Design Activities Fultz Landfill Site Byesville. Ohio Final Work Plan
https://semspub.epa.gov/work/05/161657.pdf

37) United States Environmental Protection Agency Ohio Environmental Protection Agency Proposed Plan for The Buckeye Reclamation Landfill Site St.Clairsville, Ohio May 15, 1991 https://semspub.epa.gov/work/05/41619.pdf

38) Superfund Record of Decision (EPA Region 5): New Lyme, Ashtabula County, Ohio, September 1985. Final report
https://www.osti.gov/biblio/6110563-superfund-record-decision-epa-region-new-lyme-ashtabula-county-ohio-september-final-report

39) Parole Hearing Scheduled for Man Convicted of Two Deaths - Dunkirk Evening Observer, Feb 13, 1989, p. 1 (newspaperarchive.com)

40) Double Murder is Reenacted - Courier express. (Buffalo, N.Y.) 1964-1982, December 12, 1980, Page 4, Image 4 - NYS Historic Newspapers and Courier express., September 01, 1980, Page 4, Image 4

41) Town Cemetery Arkwright, NYTOWN CEMETERY (nygenweb.net)
42) Route 83 and Meadows Road; Forestville, NY 14062 Town of Arkwright Chicken Tavern, Towns Tavern, Summit House - Chicken Tavern, Arkwright Summit, Arkwright, Chautauqua, NY (chautauquacounty.com)

43) History of Canadaway Creek - History of Canadaway Creek | Alberto Rey

44) More than 100 residents sue Arkwright project developers - SEP 26, 2019. John Whittaker - More than 100 residents sue Arkwright project developers | News, Sports, Jobs - Observer Today

45) Burnham Hollow Cemetery Map Address, Gps Coordinates And Phone Number Burnham Hollow Cemetery

46) Chicken Tavern -
http://app.chautauquacounty.com/hist_struct/Arkwright/ChickenTavern.html
http://app.chautauquacounty.com/hist_struct/Arkwright/Buyer-ChickenTavernMemories1979.html

47) Chadakoin River Chadakoin River - Wikipedia

48) NPL Site Narrative for MIDCO I MIDCO I Gary, Indiana NPL SITE LISTING NARRATIVE (epa.gov)

49) EPA Superfund Record of Decision Amendment: MIDCO I EPA ID: IND980615421 OU 01 GARY, IN 04/13/1992 -EPA/AMD/R05-92/196 1992
https://semspub.epa.gov/work/HQ/185289.pdf

50) Third Five-Year Review Report Pfohl Brothers Landfill Superfund Site Erie County Town of Cheektowaga, New York. US Environmental Protection Agency Region 2 New York, New York June 2016
https://semspub.epa.gov/work/02/393224.pdf

51) MOTCO, INC. SUPERFUND SITE Galveston County, Texas EPA Region 6 EPA ID: TXD980629851 Site ID: 0602673 U.S. Congressional District 14 Contact: Gary Miller 214-665-8318 Last Updated: August 2015
https://semspub.epa.gov/work/06/500014762.pdf

52) Superfund Fact Sheet – Western Processing Site Superfund Site. USEPA March 2000
https://semspub.epa.gov/work/10/500011196.pdf#:~:text=The%20Western%20Processing%20company%2C%20a%20chemical%20waste%20processing,a%20wide%20variety%20of%20chemicals%20and%20waste%20materials.

53) NPL Site Narrative for Toftdahl Drums Toftdahl Drums Brush Prairie, Washington Federal Register Notice: June 10, 1986 NPL SITE NARRATIVE (INITIAL TITLE) (epa.gov)

54) Technical Report Data - SUPERFUND RECORD OF DECISION _September 30, 1986 Toftdahl Drums, WA. EPA/ROD/RlO-86/009
REDACTED Superfund Record of Decision, Toftdahl Drums, WA. (epa.gov)

55) Third Five-Year Review Report Pfohl Brothers Landfill Superfund Site Erie County Town of Cheektowaga, New York. US Environmental Protection Agency Region 2 New York, New York June 2016 THIRD FIVE-YEAR REVIEW REPORT FOR THE PFOHL BROTHERS LANDFILL SITE (epa.gov)

56) Town History – Cheektowaga https://tocny.org/town-history/

57) *History of Buffalo - Buffalo in Her Formative Years* by William Chazanof Buffalo in Her Formative Years (buffaloah.com)

58) History of Buffalo - The History of Buffalo: A Chronology Buffalo, New York 1600-1799 Buffalo 1600-1799 (buffaloah.com)

59) PAOLO BUST1 (Founder of Buffalo, N. Y) https://onlinelibrary.wiley.com/doi/pdf/10.1111/j.2050-411X.1976.tb00380.x

60) A Short History of Buffalo Excerpts from "A New Look at an Old Neigborhood: Historic Homes of Bufalo's Linwood Avenue Preservation District 1820-1982." Published by the Linwood Oxford Association, Buffalo, New York, in 1992 Editor, Susan M. Pollack - A New Look at an Old Neigborhood (buffaloah.com)

61) Fourth Five-Year Review Report for Old Mill Superfund Site Rock Creek, Ohio September 2011 – USEPA FOURTH FIVE-YEAR REVIEW REPORT (SIGNED) - OLD MILL (epa.gov)

62) Explanation of Significant Differences Toftdahl Drums, Brush Prairie, Clark County, Washington REDACTED Explanation of Significant Differences; Toftdahl Drums, Brush Prairie, Clark County, Washington. (epa.gov)

63) Southern Pacific Rail Road – Sacramento - Sacramento Railyards - Wikipedia

64) Central Pacific Railroad Photographic History Museum - Historic American Engineering Record Southern Pacific Company, Sacramento Shops (Central Pacific Railroad Company, Sacramento Shops) (Southern Pacific Locomotive Works) HAER No. CA-XXXX - Southern Pacific Company, Sacramento Shops (cprr.org)

65) Trip Savvy - Old Town Sacramento: The Complete Guide By Betsy Malloy. Updated on 06/26/19 Old Town Sacramento: The Complete Guide (tripsavvy.com)

66) Old Sacramento Waterfront- About - Old Sacramento Waterfront

67) Superfund Site - CHEM-DYNE, Hamilton, OH. USEPA
https://cumulis.epa.gov/supercpad/cursites/csitinfo.cfm?id=0504440

68) Bowers Landfill – Wikipedia https://en.wikipedia.org/wiki/Bowers_Landfill

69) Remedial Investigation Report U« Remedial Evvestigation/Feasibilty Study/ Bowers Landfill, Circleville, Ohio DAMES & MOORE JOB NO. 008SO-03M7 November 18, 1987

70) Superfund Sites in Reuse in Ohio - Superfund Sites in Reuse in Ohio | US EPA

71) Cinquino, Michael A.; Keller, Marvin G.; Tronolone, Carmine A.; and Vandrei, Charles E. Jr. (1984) "Log Roads to Light Rails: The Evolution of Main Street and Transportation in Buffalo, New York," Northeast Historical Archaeology: Vol. 13 13, Article 6. https://doi.org/10.22191/neha/vol13/iss1/6
http://orb.binghamton.edu/neha/vol13/iss1/6 (original technical report - 1981)

About the Author

Peter Gorton is a co-owner of Brydges Engineering in the Environment and Energy - an environmental engineering and remediation firm located in Buffalo, New York. Born in Yonkers and raised in Peekskill, New York, Mr. Gorton received a BS degree in Public Health/Environmental Science from the University of Massachusetts and a Master's degree in Public Health (MPH) from Yale School of Medicine, where he studied epidemiology, infectious diseases, and environmental health. His 40-year career started in the infancy of the hazardous waste site assessment/remediation and chemical/oil spill response industry and included projects across the US and military bases in the pacific. He has authored hundreds of technical, environmental investigation, and assessment reports and has been involved in over ten chemical and oil spill responses at various locations across the country. Other academic/business publications include "Environmental Forensics for Cost Recovery at Petroleum Contaminated Sites" for the NYS Bar Association, "Microbiological Standards for Potentially Hazardous Foods" and he was a contributing Author for "Health & Safety for Museum Professionals – 2010 Society for the Preservation of Natural History Collections.

Mr. Gorton has spent a career reviewing the history of people, places, and events. He played five sports in High School and wrestled in college. Today he is an avid kayaker.

Peter Gorton is also the author of *The Boys of Cortlandt and the Iron Men of Croton*.

www.ingramcontent.com/pod-product-compliance
Lightning Source LLC
Chambersburg PA
CBHW042132160426
43199CB00021B/2884